A DEVELOPMENT DIALOGUE

T0275571

A DEVELOPMENT DIALOGUE

Rainwater harvesting in Turkana

Adrian Cullis and Arnold Pacey

Intermediate Technology Publications 1992

Practical Action Publishing Ltd
25 Albert Street, Rugby, CV21 2SD,
Warwickshire, UK
www.practicalactionpublishing.com

© Intermediate Technology Publications 1992.

First published 1992\Digitised 2013

ISBN 10: 1 85339 116 6
ISBN 13 Paperback: 9781853391040
ISBN 13 Hardback: 9781853391163

ISBN Library Ebook: 9781780442006
Book DOI: http://dx.doi.org/10.3362/9781780442006

All rights reserved. No part of this publication may be reprinted or reproduced or
utilized in any form or by any electronic, mechanical, or other means, now known
or hereafter invented, including photocopying and recording, or in any information
storage or retrieval system, without the written permission of the publishers.

A catalogue record for this book is available from the British Library.

The authors, contributors and/or editors have asserted their rights under the
Copyright Designs and Patents Act 1988 to be identified as authors of this work.
Since 1974, Practical Action Publishing (formerly Intermediate Technology
Publications and ITDG Publishing) has published and disseminated books and
information in support of international development work throughout the world.

Practical Action Publishing is a trading name of Practical Action Publishing Ltd
(Company Reg. No. 1159018), the wholly owned publishing company of Practical
Action. Practical Action Publishing trades only in support of its parent charity
objectives and any profits are covenanted back to Practical Action (Charity Reg.
No. 247257, Group VAT Registration No. 880 9924 76).

CONTENTS

ACKNOWLEDGEMENTS

Many individuals and agencies were instrumental in the development of the work reported in this book and I thank them for their support. To Oxfam (UK and Ireland), the major funders of this work, I am especially grateful for continual support and their confidence in backing the efforts of the project to become self-managing. There are too many other advisers from the North to mention everyone by name, but to all of them, again, many thanks.

In Turkana, many individuals were key actors in the development of the work, particularly the project staff involved in research: Pius Chuchu Nakonyi, Simon Barasa Masibo, John Munyes Kiyonga, Ambrose M. N'dirangu and Joshua Ewoi; and Margaret Ayanae and Arupe Lobuin, the team led by Cathy Watson which researched Turkana women. I would like to thank them all for their assistance, sound advice and innovative ideas.

However, nothing would have been possible without the tolerance of the pastoralists with whom we were working. In particular I would like to thank three elders whose guidance was of especial help: Ekitela Eleman, Akorot Lauren and Eyenae Ekaran. To them, and all their colleagues, I am indebted for helping me to learn.

Finally, to my co-author Arnold Pacey and the many invisible contributors to this book, the editors, typists, colleagues and friends without whose support this book would not have been completed, my heartfelt thanks.

Adrian Cullis
January 1992

PREFACE

In a sense this book closes one chapter and opens another in the history of Intermediate Technology. It is about a process of change nurtured by one agency worker in one programme in northern Kenya: in these pages is charted the progress from a demonstration rainwater harvesting project to a participative self-development project managed by pastoralists. Through its close involvement in this process Intermediate Technology, however, has been learning: it has been learning to listen to the demands of the producers with whom it works. The previous emphasis on technology transfer is giving way to a participatory technology development process in which the producers set the development agenda. Before, agency ideas were often imposed. Now, it is the producers' voice that is dominant.

Control of work, initiated by the agencies, has passed from expatriate project workers to the pastoralists represented collectively in a management committee. In this changeover no point was more critical than the decisions taken at a meeting about the management of the project after the expatriate left. In the words of Ekitela, a pastoralist from the Kachoda valley, recorded in the minutes of this key meeting: 'We do not know how to write reports and letters. ... What we need is a good secretary, but we need to make sure he does not become a manager – otherwise he will grow horns. The committee should be able to handle this person.'

The process of change chronicled in these pages, which culminates in the transfer of power, is something that we wish to share with others; the book has been written for this reason. This experience is a working example of what Fritz Schumacher, the founder of Intermediate Technology, alluded to when he championed the need for all of us to recognize the agenda of the poor majority. He said: 'The party's over. But whose party was it anyhow? That of a minority of countries and, inside those countries, that of a minority of people. And as the party became more swinging, an increasing number of people realized the party was not for them but that, at the same time, they were needed to keep the party going.' Unless the agencies wish to perpetuate the inequities of the old order, all of us in the development agencies must seek ways of working to the new agenda that is set by the producers: we must take sides. *Development Dialogue* shows how one person did this, and worked with the producers to increase their control over their own production system. We have learned in this process and hope that readers of this book may also increase their understanding. Intermediate Technology's new chapter has been opened but may not be written by us. That story will be related by the producers: it is their view that you should read next.

Patrick Mulvany
Agricultural Adviser
Intermediate Technology Development Group
January 1992

GLOSSARY

adakar	herding group
akai	night hut
amana	sorghum garden, shamba
awi	homestead, moveable camp
eboka	thanks payment
edodo	dried milk
ekarabon	elder, or elder's representative; team leader and extensionist in the project work
ekitela	herding territory or 'section'
ekitoi a ngikasakou	elders' meeting tree
ekwar	land by a (seasonal) river (for other specialized terms specifying different types of land, see Table 2 Chapter 2)
emacar	clan
emanikor	large or extensive group of gardens
emoja	name of plant from which fibres are obtained
emunyen	red dust used in make-up
eron	year when everything dies
Lopiar	name of year (1980), meaning 'the sweeping' (some other names of years are quoted in Chapter 3)
ngimomwa	sorghum

SWAHILI (KISWAHILI) WORDS

debe	container
fundi	craftsman or technician
jembe	hoe
karai	metal basin (for carrying soil in head loads)
shamba	plot of land or garden

ABBREVIATIONS AND ACRONYMS

DELTA   Development Education Leadership Training for Africa
EEC   European Economic Community
FAO   Food and Agriculture Organization of the United Nations
ITDG   Intermediate Technology Development Group
LPDP   Lokitaung Pastoral Development Project – see below.
NORAD   Norwegian Agency for International Development
ODI   Overseas Development Institute
TRP   Turkana Rehabilitation Project
UNICEF   United Nations Children's Fund
KSh   Kenya shilling

Note
The Oxfam-supported Turkana Water Harvesting and Draught Animal Demonstration Project is often referred to as the 'Demonstration Project', or simply as 'the project'. From January 1989, its name was changed and it became the Lokitaung Pastoral Development Project (LPDP).

1. THE CONTEXT OF A FAMINE, 1979–84

TURKANA IN CRISIS

THIS BOOK is about a people in Africa who live by herding animals – cattle, sheep, goats, and more recently camels. They depend on their animals for milk (a major part of the diet), also for meat and blood. They eat wild fruits in season, small game and also some cereals. The latter are bought or bartered or else are grown in sorghum gardens planted mostly by women during the rainy season.

These people, the Turkana, occupy an arid region in the north-west corner of Kenya, bounded by the frontiers of Sudan and Ethiopia to the north, Uganda to the west, and Lake Turkana to the east (see Figure 1). Turkana District covers an area of about 64 000 square kilometres, with a population of 177 000 recorded in 1984 as pastoralists or herders, keeping large numbers of livestock. An additional 70 000 people lived in small towns and other permanent settlements. Until recently, poor communications have isolated the area physically, psychologically and culturally from the rest of Kenya.

In the north of Turkana District, the worst crisis in living memory struck in 1980 when an epidemic of contagious caprine pleuro-pneumonia (CCPP) and rinderpest wiped out large numbers of animals. At the same time, there was a national cereal shortage and much-needed grain could not be moved into the area. Although this is a drought-prone region, drought was not a factor in the disaster, at least in northern Turkana. Rainfall at Lokitaung was 578mm in 1979 and 402mm in 1980, which was above average for both years and the rain was well distributed through the seasons, as Table 1 demonstrates.

As in the rest of the African Sahel region, nomadic pastoralism has evolved as a lifestyle adapted to the problems of coping with sparse and erratic rainfall. The rains come mainly in April and May, but there is a less reliable secondary peak in November (Table 1). At Lokitaung, the November rains failed between 1975 and 1987, giving less than 20mm precipitation in eight years out of the thirteen. Not surprisingly, a cycle of periodic drought and recovery is a significant part of the lives of the Turkana people. It is important to be clear therefore, that the 1979–80 famine was not part of that cycle but something exceptional.

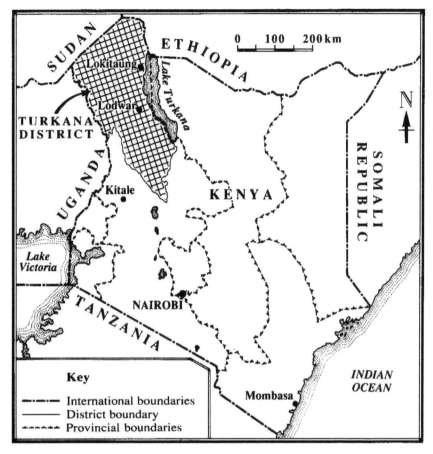

Figure 1   *Location of Turkana District within Kenya*

It seems likely, indeed, that the Turkana brought this crisis on themselves. In 1979 Turkana warriors obtained large numbers of automatic weapons as a result of the anarchy in Uganda, following the collapse of Idi Amin's regime. Armed with these weapons, the warriors then raided traditional enemies in southern Sudan and north-east Uganda, capturing large numbers of cattle and other animals. It appears that some of this livestock carried infectious diseases into Turkana District, across 'buffer' areas which would normally have separated the herds belonging to different ethnic groups. The raiding also led to deteriorating relations with neighbouring peoples, so that large areas of the borderlands became insecure and much of the vital dry-season grazing land was unusable that year. As a result, there was widespread famine in the northern part of the District, and following the intervention of the government and international aid agencies, a massive relief operation was mounted. This was

Table 1　Rainfall (in mm) at Lokitaung. The 40-year mean rainfall is quoted as 385mm per year (Hillman, 1980), while the mean for 1946–79 was 392mm (ODI 1985)

YEAR	Jan	Feb	Mar	Apr	May	Jun	Jul	Aug	Sep	Oct	Nov	Dec	TOTAL
1975	2.8	–	16.4	111.4	44.0	23.8	9.7	–	12.3	18.5	–	–	239
1976	7.0	43.6	–	33.0	–	8.2	14.8	0.8	17.8	–	47.9	–	173
1977	36.2	9.8	32.0	210.0	116.0	9.2	33.2	–	–	69.2	56.7	–	573
1978	–	60.2	126.2	19.0	–	–	6.4	8.1	51.3	13.2	10.9	111.4	407
1979	9.3	102.6	64.5	120.8	26.8	162.6	–	–	6.2	17.0	68.4	–	578
1980	23.3	–	18.6	192	85.6	1.6	–	–	–	3.6	77.5	–	402
1981	–	12.2	208.2	55.7	–	–	–	–	10.0	–	15.0	–	301
1982	12.5	–	14.5	80.4	75.8	–	5.3	2.9	–	48.9	>308	76.2	>625
1983	–	–	–	68.3	3.9	24.6	–	–	–	–	11.2	–	108
1984	*	*	*	*	*	*	*	*	*	*	*	*	<150
1985	–	–	41	144.1	21.6	–	13.4	–	–	–	9.8	3.5	233
1986	3.0	1.2	109.3	43.9	6.2	40.0	4.6	–	1.2	0.6	3.3	11.4	225
1987	0.6	44.5	26.8	86.6	71.7	24.6	–	–	–	–	50.6	8.2	314
1988	–	–	11.1	173.3	2.7	19.2	50.2	42.3	66.8	77.9	–	17.8	461
1989	2.2	119.8	32.3	64.0	72.8	–	31.6	–	16.6	–	81.4	–	421
1990	–	99.6	78.4	35.2+	–	3.4	–	–	+	+	+	+	

* Records not kept, but this was an exceptionally dry year, with total rainfall estimated at below 150mm.
+ Rains stopped 10 April; no data from September onwards.

followed by a 'rehabilitation' project executed by expatriate development workers who had little opportunity to enquire into the nature of the crisis or into local needs and indigenous institutions.

This study is about the questions of food security and development raised by the improvised programme and longer-term projects which followed. While many details quoted here are relevant only to Turkana District in the 1980s, the basic issues discussed may be relevant to other famine relief operations and to development workers in other pastoral communities. For example, the assumption behind many famine relief programmes is that once immediate needs are met, a rehabilitation or development programme will be necessary to help people restructure 'outmoded' ways of living to make them more productive. Seen at its worst, this has included the widely-held belief that nomadic pastoralists can only join the modern world if they cease their wandering lifestyle and become 'settled'. The purpose of such measures is usually conceived as making poor people richer, but wholesale destruction of their traditional institutions may well leave them less confident and with fewer coping mechanisms for dealing with adversity.

One further point concerns the role of technology in offering solutions to problems of food security and poverty. Later chapters have a good deal to say about one particular technique – rainwater harvesting – which is arguably an 'appropriate technology' for improving food production in an area with the climate, soils and human needs experienced in Turkana

District. It is important to recognize, however, that technology cannot be considered separately from the institutional or organizational arrangements necessary for its application. One therefore needs to question whether a new technology can be dealt with through existing institutions even after some evolution, or whether it also presupposes radical changes in lifestyle or organization.

THE TURKANA ENVIRONMENT

Much of central Turkana is a landscape of plains with gentle undulations but there are also mountain ranges running north–south. These mountains are formed predominantly of volcanic lavas, and are associated with the east African Rift Valley fault lines. Most of the ranges are from 600 to 900m above sea level, but the highest peak rises to about 1750m. Gulliver (1955) has suggested that above 1400m rainfall totals of as much as 750mm are common, an estimate which is supported by more recent studies of variation of rainfall with altitude (Norconsult, 1978). Rainfall is lowest near the lake shore (Figure 2), with annual averages of 400–500mm on the plains in the centre of the District.

Rainfall in the April/May peak often comes in short but very intensive storms which cannot be all absorbed by the soil. In mountains, in particular, there are considerable volumes of runoff in the form of short-lived streams. In Lokitaung itself, a threshold rainfall intensity of about 8mm per hour has been found in the hills, beyond which an increasing proportion of the rainfall enters the drainage channels (Hillman, 1980). Sometimes there are spectacular flash floods in which water levels in drainage flows may rise to above a metre in less than two or three minutes. Flash floods usually subside as quickly as they rise, and it is seldom that even the larger drainage channels flow strongly for more than five or six hours. However, in most years, floods in the rainy season sweep away an occasional homestead, and people are sometimes drowned attempting to ford swollen rivers.

Wet seasons are commonly separated by long periods without rain. During these times, weather patterns are stable with bright, sunny days. Highest mean annual temperatures in the District are recorded in areas of lower altitude including the central plain and the lake shore. Dry season temperatures seldom exceed 38°C anywhere, with mean daily temperatures varying from a minimum of 22°C to a maximum of 32°C (Hillman, 1980).

Vegetation is determined by altitude and soil moisture, with most trees restricted to mountain ranges and seasonal water courses, where higher rainfall or seasonal flows of surface-water support greater biomass production. Grasses tend to dominate the plains between the seasonal water courses. The district is classified as belonging to eco-climatic zones V and VI (arid and semi-arid), with areas of high potential restricted to the mountains (Pratt and Gwynne, 1977).

It has been widely believed by development specialists that in such an arid environment human communities must be very vulnerable in seasons when the rains fail (or partially fail). It has also been thought that the Turkana (like other nomadic pastoralists) allow their livestock to over-graze the area, and are thereby responsible for degradation of tree cover and grassland. More recent studies, however, suggest that there is no overgrazing and that degradation has not occurred on Turkana grazing lands (McCabe and Ellis, 1987). Firewood collection near permanent settlements is usually a greater problem and does lead to environmental degradation. Visiting experts who frequently spend only a short time in the area, staying in such settlements, may see disproportionately more of this. One has to beware also of applying generalizations representing all of the Sahel to particular areas. In the Sahel belt as a whole, there are certainly many areas affected by major loss of trees and grass cover. One symptom is that dust storms are now four times more frequent than in the 1960s. However, such generalizations do not fit Turkana.

Instead of accepting assumptions about overgrazing, therefore, researchers are beginning to examine more closely the causes of the periodic crises which overwhelm pastoral production systems. The Turkana have drought-avoiding strategies which enable them to survive most episodes of low rainfall, even those persisting for several years (Moris, 1988). These strategies include migration, splitting herds, hunting and the use of wild fruits and vegetables. Thus while drought plays its part in the regular cycle of Turkana life, it appears that major crises are almost never the result of drought alone. Rather they arise from the combination of several factors, which may certainly include low rainfall, but which have often also involved livestock disease and inter-ethnic warfare.

Much of the work described here was based in Lokitaung. Situated in the Lopurr Hills north-east of the District, it is a small administrative centre with a population of about 1500 Turkana together with an additional 1000 down-country Kenyans. Founded by the former colonial administration to serve isolated police posts on the northern border with Sudan and Ethiopia, it still retains an administrative function and is the base for up to 200 Administrative Police, as well as the District Officer in charge of Lokitaung Division. To the west is the Kachoda valley, which is an important grazing area used by herders.

In late 1979, townsfolk in Lokitaung became aware of losses of livestock throughout much of the northern part of the District. The numbers of animal deaths were alarming, and visiting herders told of some left completely without stock, whilst others had retained so few animals as to be unable to survive as pastoralists. For example, two herders reported

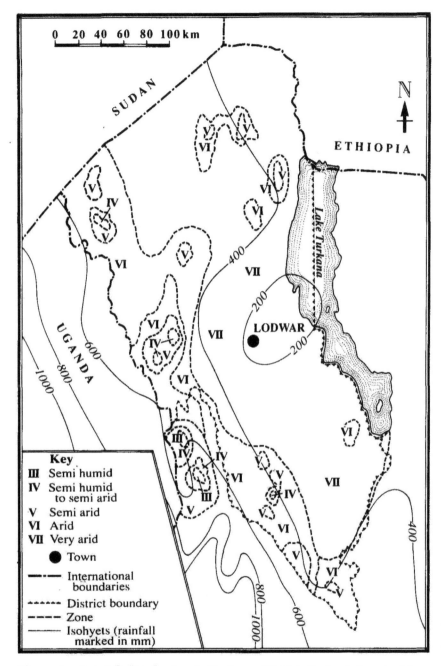

Figure 2 *Rainfall distribution in Turkana District. Isohyets (marked in mm) indicate low rainfall along the lake shore, especially around Lodwar, and higher rainfall in the west.*

that their total joint livestock holdings were now only five cattle and fewer than ten goats. Visitors said that animals were dying of diseases for which there was no cure. The rapid and dramatic spread of these diseases became more apparent when herds closer to Lokitaung were affected and decimated.

The famine years are appropriately named *Lopiar* by the Turkana, meaning 'the sweeping', due to the number of animals swept away by disease. According to herders from various parts of the District, outbreaks of disease occurred first in the far west of Turkana, and then spread eastward until the lake shore was affected. Livestock herds in southern Turkana escaped infection and therefore retained large numbers of animals throughout the crisis.

Interviews with groups of herders have enabled comparisons to be drawn between *Lopiar* and other crisis years in the 1960s and 70s. Eight herders in Kachoda, for example, record that during *Lopiar* they lost 2120 goats and sheep between them, with losses from individual herds averaging 265 animals. In two previous periods of crisis in 1961 and 1971, average losses per herd were 160 to 170 animals. The figures are in themselves unlikely to be accurate, but data collected by the Kenya Rangeland Ecological Monitoring Unit between 1977 and 1981, confirm that record numbers of animals were lost during *Lopiar*. These data indicate livestock losses in northern Turkana as 90 per cent of all cattle, 80 per cent of small stock (mainly goats), 40 per cent of camels and 45 per cent of donkeys (Hogg, 1982).

The earliest aid delegation into Turkana following the declaration of emergency thought that contrary to evidence cited here, the famine was due to drought. Their report pointed out, however, that the northern areas of the district were 'particularly vulnerable as a result of heavily armed raiders in Uganda and southern Sudan, many of whom were armed with sophisticated weapons, including automatic rifles' (EEC, 1980). Later, another team stressed the vulnerability of the pastoral economy to recurrent drought, but also identified raiding and insecurity as significant 'drought-related factors', by which some Turkana were effectively denied their herds (ODI, 1985). It is important to note that this report comments that the 'balance of power among different pastoral groups appears to have been upset by the introduction of modern weapons ...' (ODI, 1985).

Other researchers record *Lopiar* simply as a 'severe drought resulting in the massive destruction of livestock' (Hogg 1982), and this line has also been taken by the Turkana Rehabilitation Project (TRP), the major famine-relief effort launched with European Community funds.

It is not correct, however, to describe all raiding and consequent insecurity in terms of 'drought-related factors'. There is no evidence to suggest that grazing is so scarce in the region that different pastoral groups have no alternative but to fight for resources in dry years. The point seems to be that drought is politically neutral, and to present it as the cause of a crisis avoids blaming national governments or district administrators for failure to control the security situation. It also avoids the need to identify and remedy other factors which may encourage raiding – factors that might include chronic poverty, alienation from national institutions, and trading in weapons.

It was discussions with herders which provided many of the clues as to what really happened in *Lopiar*. They confirmed that immediately before the crisis grazing was plentiful and they said that many animals which died were in good condition a few days before. This is not consistent with the effects of drought, and tends to confirm that it was disease which caused most livestock deaths. However, it was not at first clear why the Turkana herds suffered such a sudden epidemic, though eventually a chance remark about cattle obtained by raiding indicated what probably occurred.

Not all the details of the story could be corroborated, but the increase in raiding following the introduction of automatic weapons was clearly a major factor (Turton, 1989). There is also an earlier parallel, in that the introduction of rifles into East Africa at the end of the last century resulted in a temporary breakdown of a relatively stable 'culture of raiding'. The consequent increase in stock theft contributed to the spread of rinderpest from 1889 onwards. It seems that ninety years later, the same pattern was repeated.

RESPONDING TO *LOPIAR*

Within a period of only a few months of the first animal deaths, many herding families in northern Turkana were finding it increasingly hard to feed themselves. It is reported that certain elders, overcome by the loss of their livestock, chose to commit suicide rather than move away in search of food. Many herders in the Division moved to the permanent settlements at Kaalin and Lokitaung. Animals which survived were left to multiply in composite herds under the care of particularly skilled herders, enabling viable herds to be rebuilt some time in the future.

People who lived in Lokitaung during this time witnessed the swelling size of the town as increasing numbers of herding families arrived in search of food. One member of the local Salvation Army staff had to accommodate his widowed sister's family, his brother's family, two aunts and several young cousins. In addition, visitors were regularly received, and consequently meals were often stretched to between fifteen and twenty people. Faced with this kind of invasion, salaried workers not surprisingly insisted that all children attend school, to benefit from the free school meals. Primary school attendance soared, and the resultant overcrowding encouraged mission personnel in Lokitaung to open an additional eight nursery or primary schools in and around the town. Adults, too, particularly women, were forced into income-generating activities, including the collection and sale of berries and firewood, charcoal and beer production, casual employment and prostitution. Increasingly desperate attempts were made to raise money by the sale of clothes and jewellery. Many of those arriving in Lokitaung in early 1980 were clearly starving.

Despite early requests for the declaration of an emergency, there were no official moves, and it was left to the Catholic Diocese of Lodwar to mount a relief operation. In Kakuma and Lokitaung, Catholic sisters opened centres for severely malnourished children which offered medical attention and prepared food. Local efforts of this kind undoubtedly saved lives, but the distribution of food only increased the numbers of herders coming to the main settlements, and the situation was rapidly worsening.

Fortunately, at this point, a local Member of Parliament alerted the international press and, faced with widespread media coverage, the authorities finally declared a state of emergency. They had been reluctant to take this step, partly for reasons of international reputation, but also because of a widespread shortage of cereals throughout the country, which meant that there was no grain which could be sent immediately into Turkana.

The Diocese of Lodwar now became involved in relief on a larger scale. Most of the food was supplied by Catholic Relief Services, which imported grain by ship to Mombasa and then by rail to Kitale. Lorries brought the food 300km to Lodwar from where it was distributed to the northern settlements. Due to inadequate transport, poor infrastructure and severe shortages of staff, food distribution was limited to towns and other settlements, and consequently even herding families still retaining small herds were attracted to the settlements. As a consequence of the rapid increase in numbers of people settling in and around the food distribution centres, service facilities were completely overwhelmed. In particular, inadequate water supplies resulted in rapid deterioration in hygiene standards, and shortly, cholera broke out. Despite gallant efforts from mission and government health workers, outbreaks occurred in all the settlements in Lokitaung Division, and many people died.

Following the declaration of a state of emergency, the Government of Kenya sought international support to fund a major relief programme. A European Community delegation toured the northern part of the district in early June, visiting the feeding centres and meeting mission personnel. The delegation reported that 27 000 people were completely destitute, and numerous others had lost large numbers of livestock (EEC, 1980). They commented that the Turkana experience a 'traditional cycle of overbreeding livestock' leading to overgrazing, and famine following periodic drought. There was then 'partial replenishment of livestock by raiding', with many animals taken from 'other tribes' (EEC, 1980).

The delegation recommended against the rehabilitation of herders solely by replacing lost animals. Instead, it proposed a more broadly-based rehabilitation programme, including measures to combat desertification and improve marketing of surplus animals. In this way, the 'classical famine relief intervention, which resulted in the creation of permanent destitutes' could be avoided (EEC, 1980). Had it been realised that the Turkana pastoral economy was potentially as robust as ever and that livestock diseases were to blame, the response might have been different.

In November 1980, a memorandum of understanding between the European Community and the Kenya and Netherlands governments was signed which established the Turkana Rehabilitation Project (TRP), a programme that was to supply 5000 tons of food each month. Its objectives were the emergency feeding of destitute people and the rehabilitation of the land and population, partly through food-for-work, but partly also by the provision of viable flocks to selected families.

When the TRP eventually became operational more than a year after the crisis began, mission personnel continued to play key roles. The first TRP manager, for example, was a diocesan priest who had worked in Turkana for many years, and who managed the project until early 1983.

In early 1981, TRP recruited Standard VII leavers from local primary schools as 'site' (camp) managers and adult education teachers. In Lokitaung, more than forty staff were appointed within two days and, with minimal instruction, were set to organize the first registration of destitutes and to begin to organize adult literacy courses.

Some attempt was made to divide the famine camps into smaller units of forty to fifty families as had been suggested by TRP and to identify camp leaders to assist in registering destitutes. As TRP appointed more staff, 'site facilitators' were made responsible for the distribution of food to two or three camps. Later, European volunteers were recruited to coordinate food distribution and implement a programme of food-for-work while also supervising the recently recruited local staff.

Once the registration of the destitutes in the settlements was complete, new arrivals obtained emergency food rations without waiting for the weekly distribution. It was clear from the physical condition of many of them that without relief food they would have starved. However, the distribution of food in the settlements rapidly became associated with widespread abuse and many people who were not destitute registered for relief. As it became clear that it would be impossible to prevent this, less effort was made to screen camp lists for the families of local traders, teachers and government officials who were not entitled to relief. Instead, efforts were concentrated on ensuring that adequate food went to the poorest and least powerful people in the camps, usually the elderly or disabled, and members of women-headed households.

The numbers of destitutes recorded rose to more than 80 000 in 1982, but checks indicated that registers were often inaccurate by as much as 20 to 30 per cent. It is therefore unlikely that more than 50 000 herders were made destitute by *Lopiar*. However, there were good arguments against screening too strictly to ensure that food went only to those who need it, the most important of which was that the distribution of food to some people who still possessed animals avoided the need for stock to be slaughtered and permitted the more rapid rebuilding of herds (Hogg, 1982).

Families were also helped to acquire new animals through barter with a variety of sources from which to obtain relief aid, including TRP, the Red Cross, Catholic Relief Services, Catholic missions, the Salvation Army, and government agencies. Therefore, by the careful management of food supplies at family level, it was possible for herders to purchase animals at rates comparable with traditional barter rates. Herders in Kachoda, for example, record the exchange rate at 2 to 2.5 debes, or 40–50 kilograms of grain for a goat. This is equivalent to about KSh100/– to KSh125/– per goat. By contrast, where goats were purchased for cash, prices were higher at around KSh175/– to 200/–. Herders seeking to restock evidently benefited from the widespread adoption of traditional barter rates such as prevailed for sorghum, rather than observing cash prices.

FROM RELIEF TO FOOD-FOR-WORK

At an early stage it was realised that people had exaggerated expectations of TRP. The activities of TRP in 1981 certainly gave the impression that life was to be transformed. In Lokitaung, staff houses, stores and dispensaries were being built, as also at other smaller centres such as Loarengak, Kachoda, Kaalin and Kaikworr. In addition, TRP improved local water supplies by drilling boreholes and installing windpumps and handpumps. A large fleet of vehicles also appeared, including twenty lorries, fifteen tractors, some water bowsers, and several smaller vehicles and motorbikes (ODI 1985). For local people, the change from living in a remote district largely cut off from the rest of Kenya, was both sudden and dramatic. It is perhaps not surprising that people later expressed disappointment in TRP, which had seemed to promise so much.

In 1980–81, the bulk of the emergency food was provided by Catholic Relief Services, the European Community, and the Netherlands Government. In March 1982, however, the World Food Programme signed an agreement to supply food under its emergency programme. During the remainder of 1982, this involved sending 40 000 tonnes via Kitale, a level of support which could not be sustained. There was pressure, therefore, to reduce the numbers receiving food aid by distributing it via a food-for-work scheme. Local leaders and MPs launched a campaign against any reduction in aid. There were also several incidents where angry crowds threatened volunteers whilst making demands for food, but food-for-work was gradually phased in.

Despite early concern that food rations were to be immediately and drastically reduced, the Turkana often record, with wry humour, their first experiences of quantified food-for-work. They point out that the administrative procedures were so poor that it was possible to continue receiving large amounts of food for limited work. In the early stages, much of the work involved the construction of earth bunds for the retention of rainwater to encourage improved crop production in 'gardens', or microcatchments for the establishment of trees. However, for most of those involved, the object of the work was not appreciated. It seemed to them that all that was required if they were to receive food was the rapid heaping up of large earth structures. Brushwood was sometimes buried under some bunds to make them look bigger, and surveyors sent to measure the work done were sometimes intimidated.

During 1982–3, it was found that many men left the work sites, and that increasingly, it was women and children who were carrying out the food-for-work tasks. Hogg (1982) cites a place on the Turkwel River, where out of 759 adults, 520 were women. It seems that the missing men were herding, or seeking waged employment. Large numbers of herders aimed to remain within the pastoral system, and were temporarily 'off-loading' women and children until they were able to build up their herds. Some purchased animals with surplus grain, while others obtained stock by raiding neighbouring tribes in southern Sudan. It is reported, for example, that as early as 1981 several raids involving up to 3000 warriors took place. Certainly, without raiding it is difficult to explain the rapid decline in the numbers of destitutes in certain camps. By contrast, there were reports of families so dependent on food-for-work that they denied themselves opportunities to travel and meet relatives, preferring to work seven days a week rather than forfeit their places in the camps.

The increasing powers placed in the hands of relatively young and often poorly supervised site facilitators allowed the camps to be increasingly manipulated for personal gain. This said, however, it has to be pointed out that there was considerable dissatisfaction within TRP's workforce at the time and, in 1983 alone, more than 30 per cent of site facilitators and adult education teachers left TRP for other work. The reasons given include low wages, inadequate training, limited career opportunities, and the reduction of allowances. The reduction of food aid reaching the camps may also have been a contributing factor.

Despite all these difficulties, large numbers of people remained in the camps, and other strategies had to be found to reduce their dependence on food aid. As it was almost impossible for area co-ordinators to deny people the right to work if maize was in the stores, the World Food Programme and TRP management decided to wait until several months after the rains and then stop the supply from Kitale. In this way it was hoped to force people to return to the pastoral sector. Thus in July 1983, the regular shipments of more than a thousand sacks of maize a week were suspended and, for three weeks, no maize was imported. The regional stores quickly emptied and the food-for-work programme came to an abrupt halt. The result was roughly as intended. Increasing numbers of people abandoned the camps and returned to their herds. However, some people remained and it became apparent that the very fact of feeding destitute persons had effectively created and maintained a new class of 'stockless' Turkana. In previous crisis years, these herders would probably have died or else become absorbed by a neighbouring agricultural tribe. TRP had offered an alternative and was now looked to as a major source of income. The effect of the stoppage in maize supplies on this section of the population was one of considerable hardship.

PROGRAMME EVALUATIONS

The reduction in food-for-work opportunities continued through 1984–5 as the import of grain was cut from 824 to 450 tonnes a month. The resultant decline in beneficiaries from 1982 onwards was from 40 000 to perhaps 11 000 in mid-1985 (ODI, 1985). There is evidence that most of the people no longer receiving food returned to the pastoral economy, which indicates that TRP largely succeeded in meeting its stated aims.

Thus an evaluation carried out by ODI (1985) regarded the establishment of the system of food transport, storage and distribution as 'a notable success'. It is interesting to note that evidence offered in the ODI report to support this claim included the passing unnoticed of the 1984 drought, since food aid was already being distributed before it began. Rainfall data (Table 1) indicate that the effects of this drought could otherwise have been serious. The herders themselves agree that TRP's relief operation was effective and that many thousands would have starved without it.

Whilst tribute was given to TRP's handling of the relief operation, the ODI report identified a number of problems, including the settlement of large numbers of destitutes in famine camps. There were no employment or income-generating opportunities for so many people, and it was clearly not desirable to prolong their dependence on food aid. Another problem associated with the camps was the destruction of tree cover in their vicinity as people collected firewood. This prompted the suggestion that large concentrations of destitutes should be avoided in any future famine. If possible, herders should be encouraged to remain with such of their animals as survived a crisis (ODI 1985).

The ODI report was highly critical of the food-for-work aspect of the programme, as was another evaluation by Schwartz, Schwartz, and van Dongen (1985). According to the latter, TRP management placed 'unjustified confidence' in the rainwater harvesting work which the people were asked to undertake. This was an 'imported technology', and would have 'little or no positive effect on the range areas' used by the pastoralists.

Several detailed proposals for the future of TRP's work came out of the Schwartz study. In particular, a re-orientation to the needs of pastoralists was recommended. The report identified the need for improved training of pastoralists and the establishment of veterinary and livestock development centres. It was argued that the training should be developed out of a proramme of research into appropriate range management, animal husbandry, and water-resource development techniques.

These evaluations by ODI and the Schwartz team were disappointing partly because both failed to link the outbreak of livestock disease with Turkana raiding. It is important that this link be made so that the extraordinary nature of the *Lopiar* crisis is fully understood. It becomes clearer then that the Turkana economy is extremely robust when faced with normal recurrent droughts. An appreciation of the role of modern weapons in the crisis might also have suggested alternative policies, including peace initiatives with neighbouring ethnic groups, improved border security and accelerated restocking.

A second shortcoming in the studies was a failure to make recommendations for rehabilitating ex-herders who had been unable to re-establish viable flocks. Before the 1960s, herders made destitute by whatever set of circumstances starved to death, migrated out of the region, or supplemented their livelihoods by hunter-gatherer activities or lake fishing until they were able to rebuild their herds. Following *Lopiar*, TRP's relief operation provided destitute herders with a more attractive alternative, and whilst some have been able to re-establish themselves as herders, others have not. TRP has therefore unwittingly added to the number of stockless ex-herders, a social group which first emerged with the introduction of relief operations in the 1960s. Stockless ex-herders are found in each settlement in Lokitaung Division and casual observation indicates the extreme distress of their condition.

To conclude, then, the two major evaluations of relief and rehabilitation efforts in Turkana, both completed in 1985, raised questions about long-term development in the region, not only by what they explicitly recommended, but also through their omissions. Some of these questions will be pursued in the next chapter.

2. HISTORY AND DEVELOPMENT TO 1984

INTERPRETING PASTORAL ECONOMIES

Within the Sahel region of Africa, pastoralists and agro-pastoralists have rarely been incorporated into the political and economic life of the nations to which they nominally belong. Furthermore, current development policy, often supported by international funding, has failed to address local concerns and priorities. The result is a sense of alienation and mistrust as governments seek to 'modernize' traditional production systems.

In an area where many people lead a nomadic life for at least part of the year, the development programme tendency to focus on permanent settlements inevitably erodes local institutions and adds to the sense of alienation. So also may the concentration of famine relief on a number of large camps which usually persist for longer than intended. It needs to be understood that pastoralists have coped with periodic crises in the past precisely because they have *not* been closely tied to fixed homes.

Ironically, as the habit of equating 'modernization' with 'settlement' has become stronger, increasing numbers of research studies have produced evidence for the complexity and wisdom of traditional pastoral and agro-pastoral systems (Swift, 1981). There is a growing body of opinion which holds that policy planners should consider in greater detail the advantages afforded by tried and tested methods. There is also an increasing recognition of the importance of strengthening low-level pastoral institutions.

There are several reasons why it is so often assumed that the future of nomadic pastoralism can only lie in settlement. One is the assumption that modern services, notably education, can only be satisfactorily provided in towns and permanent villages. Another is the idea that a growing population of pastoralists with increasing herds of livestock is already over-exploiting the environment. Linked to this is the belief that as overgrazing and desertification worsen, periodic crises such as the *Lopiar* famine will become more frequent and more devastating. Thus the only way to avoid future famine problems might seem to be through developing alternative forms of livelihood such as irrigated agriculture. But while there clearly must be a limit to the numbers of livestock and people an area can support, desertification has more often been accelerated by activities associated with permanent agriculture or small towns than with pastoralism itself. Thus one United Nations agency has argued with respect to the Sahel, that settlements 'can exacerbate ecological problems', and hence that nomadic pastoralists 'should not be forced into a sedentary lifestyle'. The argument continues that pastoralists will settle on their own initiative, 'when they sense it to be advantageous and workable . . . their judgement, based on their intimate interaction with nature, is usually sounder than that of outside technical experts.' (UNICEF, 1985, p.38).

In Turkana, Gulliver (1952) reported that areas which offered the best grazing were not being over-exploited, because seasonal movements of stock to other areas were successful in conserving them. Twenty years later, Henrikson (1974) noted claims that there was wide-spread deterioration of pastures, but commented that in a purely pastoral production system herders would drop out in a crisis so that a balance between population and environment was maintained. Thus long-term cumulative degradation was avoided. In Turkana, he thought, famine relief had provided destitutes with an alternative lifestyle and in some cases made possible the rebuilding of herds. It was the resulting increase in herders that Henrikson believed could lead to eventual collapse of pastoralism through the destruction of the resource base. He was by no means alone in this, and much development thinking about Turkana has been orientated around the need to identify economic alternatives to pastoralism.

In 1979, the government of Kenya launched its Arid and Semi-arid Land Programme, which was designed to give higher priority to marginal areas. Unable to bear the cost of the programme itself, the government sought support from western governments. A number of bilateral agreements were made which linked western donors with specific districts, including one linking Norway with Turkana in 1980. This agreement resulted in substantially increased funding from the Norwegian agency NORAD through the establishment of the Turkana Rural Development Programme. During 1979, four teams were sent to Turkana to carry out a variety of studies. The first wrote that 'land degradation and environmental destruction is getting worse. With increasing numbers of livestock, the system is very fragile'. Most experts feared a future of recurrent droughts and tragedies for animals and humans if no major change occurred (NORAD, 1979). In the light of such thinking, it is hardly surprising that the mission recommended reduction of livestock numbers by encouraging increased sales of animals out of the district, and absorption of some of the nomadic population into other kinds of activity, such as agriculture or fishing on Lake Turkana.

When the NORAD programme was reviewed in 1988, it was recognized that alternative economic activities had largely failed and remained uncompetitive with pastoralism. The review team was critical of many of the development initiatives taken, which they regarded as extravagant. It was suggested that not only had resources been denied to pastoralists, but that by concentrating these resources in the settlements, much had been done to the detriment of pastoralism (Sorbo *et al.*, 1988). Environmental degradation around settlements was mentioned as a major concern. In contrast, the review quoted research which had found that 70 per cent of the district could support more livestock. Failure to encourage pastoral development was therefore brought into critical perspective.

The review concluded that future development initiatives in the district should seek to support pastoralism in particular by seeking to reduce the risks faced by pastoralists; by support for local institutions; by encouraging local participation in development planning; and by reducing the flow of resources into the district to a level appropriate to local administrative capacity (Sorbo *et al.*, 1988).

This change of view within the NORAD programme is highly significant, but to investigate further the conflicting claims made about the Turkana pastoral economy and its potential for development, it is necessary first to look in more detail at the history of the Turkana people and of programmes designed to benefit them. Then second, it is important to understand Turkana institutions, their role in coping with seasonal change and periodic crisis, and their future potential. This latter topic is dealt with mainly in the next chapter.

HISTORICAL PERSPECTIVES

The Turkana became differentiated from other ethnic groups some 250–300 years ago, and were well established in the western part of their present territory soon after 1700. By 1900, as a result of raids and conquests, they occupied a larger area extending to the lake shore, with a population in excess of 30 000 (Lamphear, 1976). Their success was due to good organization and contact with iron workers in Labwor in what is now eastern Uganda, who provided metal for spears and tools.

European traders, explorers and ivory hunters began to appear from the 1880s on. Most commented on the friendliness of the Turkana, with whom they traded, but one Italian-led expedition killed 300 Turkana. With increasing interests in the region, both the British and Ethiopians claimed the region. The first major British expedition was in 1898.

Following a new rinderpest epidemic in 1908, the Turkana sought to rebuild their herds by raiding other groups. This led the British to launch a military campaign aimed at controlling them, followed by an attempt to collect a 'hut tax'. Little was achieved, however, until in 1915, a British force subdued the southern Turkana after killing 400 warriors and capturing large numbers of stock. Two years later, after the northern Turkana had raided peoples to the south who were under British protection, a large force was sent against them. After fighting two pitched battles, organized

Turkana resistance was crushed in 1917. It is estimated that during 1916–18, the Turkana lost 250 000 head of cattle to the British. Certainly, the herders in Lokitaung Division had very few animals, and herds did not recover until after 1930. The British had conquered and impoverished the Turkana people, but did not know what to do with the territory which they regarded as desert. Lokitaung became the first administrative centre for the District in 1927, but this function was later moved to Lodwar, in central Turkana. The British restricted travel, maintaining it as a closed area and cutting the Turkana off from development elsewhere. Particularly harsh measures imposed on the Turkana included a ban forbidding them to wear European clothes, notably in the town of Lodwar. Henrikson (1974) also notes that the Turkana were denied any educational facilities, and he argues that this alienated them from the possibility of development. Certainly, the lack of schools meant that, after independence, officials had to be brought in from other districts, and many of them regarded the Turkana as backward and uncivilized. As will be discussed below, this has had repercussions on development initiatives.

Hogg (1986) records that 'as early as the 1930s the then British administration distributed famine relief'. It is likely that Hogg learned of this relief operation in the colonial records, maintained at Lodwar, which perhaps explains why no mention is made of the loss of more than 250 000 animals to the British in 1916–18. It is possible that, although the Turkana had rebuilt viable herds by the 1930s, the pastoral economy remained more vulnerable to crisis than before.

Immediately before independence in 1960–61 the Turkana pastoral system was again overwhelmed. The British authorities mounted a large relief operation to feed over 10 000 people who were seeking help in famine camps (Henrikson, 1974). Reluctant to be drawn into a protracted feeding programme, the District Commissioner requested the support of missionary societies to co-ordinate the relief operation. Mission staff from the Catholic diocese of Nakuru were sent into Turkana, and established feeding centres at Lorugum, in the central-western part of the district. Within the next few years additional centres were established throughout northern Turkana, and primary schools and dispensaries opened. Some of the first priests and sisters who came into the area have remained working in the district until the present time. Other missions also became involved, including the Africa Inland Mission and the Dutch Reformed Church of East Africa. In Lokitaung, the Salvation Army established a mission in 1961. The mission initially responded to the famine by mounting a relief operation, but later built several primary schools. It is interesting to note that all the first intake were boys, and it was not until later that girls were allowed by their parents to attend school.

Even today, many pastoralists resist attempts to persuade them to send girls to school. Anxious to rehabilitate destitute herders, the administration encouraged some to move to the lake and engage in fishing. Henrikson (1974) refers to the posting of a fisheries officer in 1961 who was under instructions to carry out a simple fisheries survey and experiment with the production of dried fish. A more novel approach to rehabilitation was implemented at Kaalin, where the District Officer restocked herders with small numbers of animals. It is reported by local people that adults were each given up to twenty sheep or goats together with some provisions, and were then sent back into the bush. Unfortunately there is little documentation concerning the success or otherwise of the scheme. However, at least two of the herders who were restocked at this time maintained viable herds until *Lopiar*, and they and others speak of the great value of the restocking exercise. This novel and extremely popular method of rehabilitation was repeated by Oxfam, following *Lopiar* (Burke, 1987). It could be argued that, from the Turkana point of view, restocking is long overdue as a form of 'compensatory payment' for the losses experienced under the British.

THE CULTURE OF RAIDING

In the histories of most pastoral peoples, the raiding of cattle from other ethnic groups is a continuing theme. In Turkana, it brought expansion and consolidation during the nineteenth century (Lamphear 1976), but led to disaster later – first through its role in the spread of rinderpest after 1889, then in provoking the British effort to 'control' the Turkana from 1916, and most recently by introducing the diseases associated with *Lopiar* in 197t–80.

One unplanned benefit arising from the food-for-work programmes of the early 1980s is that the distribution of grain provided people with a means to rebuild their herds by purchase rather than by raiding, and it perhaps needs to be recognized that as long as pastoralism is practised, such means of rebuilding herds after periodic crises will be necessary. That is, any development programme among pastoral people needs to pay attention to means of making new stock available, through markets or otherwise.

Consideration of raiding and the occasional convening of bands of armed warriors not only raises questions about stock replacement and peace-making with other ethnic groups. There are also questions concerning the possible social function of raiding. Turkana political institutions were traditionally based on a generation-set system through which groups of young men of similar age were formed into fighting units. This meant that, despite the decentralized lifestyle of the herders, it was possible to gather together large numbers of warriors when necessary. Until about a century ago, clashes between these warriors and neighbouring ethnic groups did little damage. Few people were killed or injured, and while many livestock might be captured, it was usually possible to make peace and 'bury spears' relatively easily. However, the advent of firearms in the nineteenth century and more recently, of automatic weapons has made raiding steadily more lethal. Moreover, in recent times, neighbouring states have been engulfed in civil war and unrest, and the Turkana have had every reason to feel that they need to be able to defend their lands and herds – and the Kenya government has encouraged them to do so through the arming of the 'home-guard' units.

However, in the long term one must hope for a return to conditions which make raiding and the organization of warrior bands unnecessary. Certainly, one cannot think of 'development' as involving the expansion of herds by stealing other people's cattle. However, we need to recognize that the culture of raiding has probably been of considerable significance in Turkana life, especially for men and youths. It has given the latter opportunity for comradeship in risk-taking, and for competitiveness with regard to courage and skill. Such experiences seem to be important for some men in almost all cultures to enhance their sense of meaning in life and express their masculinity.

In Turkana, the skills and challenges necessary to keep a herd alive through a difficult season provide other outlets for some of the same impulses. Even so, if development processes were to displace an institution as prominent in traditional life as raiding and the generation-set organization which supports it, thought would have to be given to the social consequences. It would be important to recognize the dangers of leaving some individuals without a role which once gave meaning to their lives.

Much attention is now rightly given to the role of women in development, but it needs to be appreciated that there is sometimes a problem of men's roles too, which left unaddressed, can leave some men rootless and alienated so that they resort to drink, petty crime, or 'gangland' activities – or else seek a continuation of the old warrior culture by joining the armies fighting civil wars in this region. In other words, there is a need for development to create new roles for men in order to provide challenges and satisfactions comparable with what was formerly associated with the culture of raiding.

The point has been made by Boserup (1981) that it suited former colonial regimes to have a reservoir of rootless men displaced from their traditional society roles, because they then formed a supply of cheap labour for mines and plantations. One difference, therefore, between modern and colonial thinking about development ought to be recognition of the problems of men's roles in society.

FISHING AND IRRIGATION

The first significant efforts to introduce new economic activities into Turkana District began in the 1960s, and included three kinds of programme: fisheries in Lake Turkana, conventional irrigation, and water-spreading (or 'spate irrigation'). The general assumption behind much of this work was that development would mean the abandonment of the nomadic lifestyle, and that people would become 'settled'.

Fishing in Lake Turkana was promoted after the famine of 1961, when some destitute herders were offered the opportunity of resettlement at Kalakol in order that they might be helped in becoming fishermen.

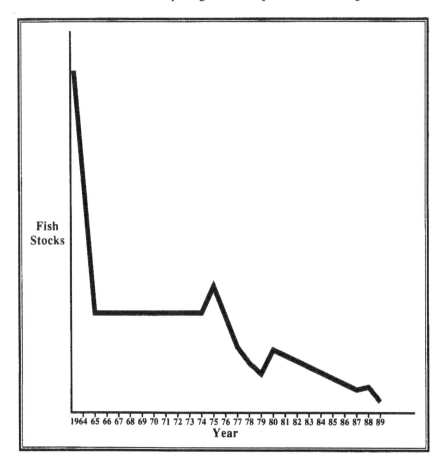

Figure 3   *Estimates of fish stocks based on experience of declining catches in Lake Turkana. The estimates were made by three Turkana fishermen using pebbles to indicate quantities. Formal data show exceptional catches in 1976, then a steady decline from 5000 tons in 1976 to 600 tons in 1985. (Source: Watson 1990)*

The National Christian Council of Kenya, through the African Inland Mission, with support from Oxfam and other non-government organizations, provided destitute people with gill nets and canoes. Henrikson (1974) records that in 1967, more than 1700 tons of dried fish were exported. Furthermore, encouraged by the success of the programme, an additional two fishing camps were established, including one at Loarengak, fifteen kilometres west of Lokitaung. In this way it appears to have been possible to rehabilitate significant numbers of people, for Henrikson refers to the removal of 750 individuals from the registers of famine camps. Moreover, with more help from donor agencies, a fishermen's co-operative was formed to assist with marketing dried fish.

Encouraged by the large catches of the early 1970s NORAD helped pay for a tarred road partly intended to improve the transport of fish from the lake to down-country markets. A fish-freezing plant was also built, but by the time it was completed, fish catches had dropped (Figure 3). The plant is now an internationally famous white elephant, never used for its original purpose. More noteworthy, however, is the fact that as fishing declined, some families who had bought livestock with the money earned this way moved away from the lake shore and resumed their traditional pastoral lifestyle.

With regard to agricultural development, Henrikson (1974) commented that a consultant had visited Turkana ten years earlier and reported that there were possibilities for 'large-scale irrigation schemes, by means of dams' on the Turkwel river, but questioned the economic viability of any such schemes. However, one project began in 1966 with financial support from a German agency and technical aid from FAO (Sorbo *et al.*, 1988). More than 20ha were cleared by 50 families and large numbers of tree crops were planted, including citrus, dates and coconuts. Sorghum, vegetables and sunflowers were also grown. The second scheme was supported by the Catholic Diocese of Lodwar, which resettled small numbers of destitute people at the site. The settlers continued to receive food relief while they clear-felled an area of 30ha. The scheme was later twice expanded, and in 1979, more than 100ha were being farmed by 300 families (NORAD, 1979).

Following the apparent success of these two Turkwel schemes, FAO and the United Nations Development Programme (UNDP) opened a gravity-feed irrigation system at Katilu (Figure 2) in 1971. Initially the scheme was quite small (20ha), but it was later expanded to 212ha. There appears to have been considerable pressure for the irrigation scheme to produce large amounts of food quickly, for highly mechanized systems were adopted, with central management. In more recent years, other irrigation schemes have been established on southern lengths of the Turkwel and to the east along the Kerio River (Figure 2). The total irrigated area in Turkana is now nominally about 1000ha, nearly all on these two rivers, though it is estimated that the irrigated area actually in use may be less than 500ha (NORAD, 1990).

During the early 1980s, NORAD commissioned several reviews and studies of the Katilu scheme, one of which stated that 'the project had been unable to achieve its basic goal, i.e. to ensure its participant farmers with yearly incomes sufficient to cover subsistence needs ... it is a disguised famine relief camp' (Broch-due, 1983). The result was that households engaged in a variety of other subsistence or income-generating activities, including stock-keeping. Much of the burden for this work fell on women, which resulted in the breakdown of families and the emergence of a large proportion of women-headed households. Yet it was taken for granted that men were the providers, and women were allowed little influence in the decision-making processes, as they were denied full membership in the co-operative (Broch-due, 1983).

Moris (1987) suggests that irrigation schemes once enjoyed a 'privileged status' amongst policy-makers and planners in Sahelian Africa, because irrigation appeared to offer an obvious solution to drought and food insecurity. As a result, many schemes escaped 'detailed justification and local adaptation', and have been increasingly associated with high costs and poor performance. The Turkana schemes are amongst the most expensive in Sahelian Africa, at an estimated cost of US$50 000 per hectare (Moris, 1987), or US$61 240 (Hogg, 1988). With these large capital costs now more widely recognized, irrigation has ceased to be a privileged technology among aid donors, but there are still economic, environmental and social costs to be met in Turkana. One engineer commented that like many other regions in Africa, Turkana has 'borne the brunt of a lot of bad advice, bad engineering and bad intentions, and they are paying for it now' (quoted by Moris, 1987). Yet there are still plans to extend the area devoted to irrigated agriculture, following the recent daming of the Turkwel River at Marich.

THREE WAYS OF RAINWATER HARVESTING

An interesting initiative in the early 1960s which has not so far been mentioned was the introduction of water-spreading and other rainwater harvesting techniques. The pioneer site was on the Lorengippe seasonal river and the work was partly funded by Oxfam. A system of bunds was built to slow (rather than totally hold back) runoff water coming onto the site during heavy rain (Figure 4). Food crops were successfully grown in two seasons before the site was taken over for multiplying grass seed.

This was a far smaller project than either the fishing scheme or the conventional irrigation projects, yet in contrast to these, it did not presuppose the abandonment of nomadic pastoralism by the Turkana. It was also closely compatible with their existing tradition of growing some

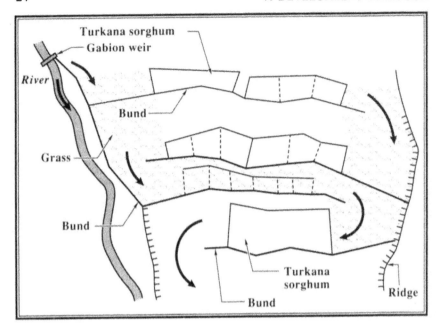

Figure 4   *Water spreading from the Lorgengippe 'sand river' in the south-
west of Turkana District. Water was diverted from flood flows in the sand
river by means of the gabion weir, and was spread over the land by the
system of bunds shown on this plan. The bunds were aligned to fall away
very slightly from contours at 0.7m vertical intervals. The largest bund was
390m long, 1m high, and had a 3:1 slope on its upstream face. The system
was constructed in March 1963; seed was sown on the plots marked for
Turkana sorghum when rain fell during the next month and a good crop was
produced. (Source: Fallon, 1963)*

crops during the wet season. Under this system sites are carefully selected
where large volumes of floodwater or runoff naturally flow during storms –
a technique which has been characterized as the use of 'simple' or
'informal' water-concentrating methods (Pacey and Cullis 1986). However,
the scale of the earthworks for water spreading was rather large, and would
have presented problems of organization and labour requirements if many
such schemes had been built.

  For almost twenty years, the Lorengippe project and one or two similar
experiments remained isolated examples of formal water harvesting
(Hogg, 1987) until the Salvation Army initiated some new work. This
followed a visit to Israel by their rural development officer who saw
rainwater harvesting in use in the Negev Desert and later suggested that
similar techniques could be relevant in Turkana. The British agency,
Voluntary Service Overseas (VSO), recruited a volunteer to develop the
idea with the Salvation Army in Lokitaung, who also visited the Negev

before arriving in Turkana in early 1978. Work done during the next two years is well documented (Hillman, 1980), and it is clear that every effort was made to incorporate rainwater harvesting into the local production system. The project sought where possible to improve the reliability of traditional cultivation. Gardeners were subsidised through food-for-work payments to develop their sorghum plots and later to plant drought-resistant browse and fodder tree species. Despite the use of simple surveying equipment and the construction of bunds along the contours, floods caused considerable damage and the sites were in constant need of maintenance. Yields were recorded at around 200kg or sorghum per hectare, but this figure is so low as to be questionable. After two seasons, there was considerable scepticism as to whether the Israeli technique could easily be duplicated in other arid areas.

At this point, it may be desirable to clarify the definition of rainwater harvesting. In its broadest sense, this term refers to any process whereby crops, grass or trees are grown in an arid area by exploiting runoff water flowing from large catchments during rainstorms onto smaller cultivated

Table 2   Local classification of rainwater harvesting sites used by Turkana gardeners (for location of sites, see figure 15, chapter 5)

Type of site (Turkana name)	Example (where found)	Description of site
ataan	Naramum (Kotome valley)	delta fan, where runoff from the mountains is checked on the plain
aroo	Lomareng (south of Kachoda)	many small streams crossing a plain where water is spread naturally
alelesi	Manalongoria (north-west of Lokitaung)	an area of fertile alluvial soils beside a river bed
ekwar	Natoo (near Loarengak)	braided stream with planting carried out on the islands
atapar	Nayanae Losenyanait (lake shore)	natural pond where runoff is retained in depressions or by sand-dune barrier
apasi	Namuruputh (near Todenyang, lake shore)	drainage of freshwater runoff impeded by saline lake water
ejem	Lomogol (north end Kachoda valley)	swampy area where major drainage channel discharges onto the plain

plots. Sometimes the catchment is an area of sloping ground adjacent to the cultivated plot and rainwater runs directly onto the plot without entering a water course. Sometimes floodwater flowing in a usually-dry channel or *wadi* is utilized.

Most of the traditional sorghum gardens made by the Turkana depend on rainwater harvesting in this general sense, and have been cited as examples of the informal use of this technique (Pacey and Cullis, 1986). Morgan (1974) showed how some Turkana gardens in the Kerio valley were situated below hillside catchments, while others were on land flooded by seasonal rivers. Interviews with cultivators near Lokitaung and on the lake shore indicate a similar distinction. It will be apparent that Turkana gardens are not 'rain-fed agriculture' but rather come within Stern's very broad definition of 'irrigation' (1979, p.13), and within most definitions of 'rainwater harvesting' (Pacey and Cullis, 1986).

It is a mistake to assume that either rainwater harvesting or irrigation must always involve earthworks or other engineering structures. The traditional Turkana approach is to use sites where rainwater harvesting is achieved by the natural shape of the land. The skill involved is largely a matter of site selection, which either depends on observing where flooding occurs in the wet season, or on noting certain plant species and soil types in the dry months. Heavy thorn barriers sometimes surround gardens, mainly to keep out livestock, but they also serve to hold back debris carried by runoff flows, thereby slowing and speading the water flow on the garden plots (ASAL, 1990). By contrast, the rainwater harvesting techniques introduced at Lorengippe and later by the Salvation Army and TRP, were based on building earthworks of various kinds to control runoff water or flood flows. These new techniques may potentially increase the number of sites on which sorghum can be grown, and are of two main kinds.

One is 'spate irrigation' or water-spreading, as tried at Lorengippe. This is sometimes known as the 'Yemeni method' and usually involves a relatively large-scale operation in which water is diverted from flood flows in a natural water-course or wadi. Low banks or bunds are built to slow down and spread the water over the cultivated land. In some versions of this technique (Figure 4), the bunds are not designed to stop the flow of water entirely and it is not 'ponded' on the land.

With the other approach, as used in the Negev Desert and parts of Sudan, bunds trap runoff water on levelled plots or 'terraces' (Figure 5), where it stays until it has soaked into the soil or evaporated. The bunds are built with spillways which allow any excess water to escape, but apart from overflows, the water is simply allowed to flood the cultivated land. In Turkana bunds are built, generally, to capture runoff water from relatively small catchments and thus minimize the risk of damage to bunds and spillways from excess flooding. In a few cases additional floodwater is diverted or captured from larger stream flow.

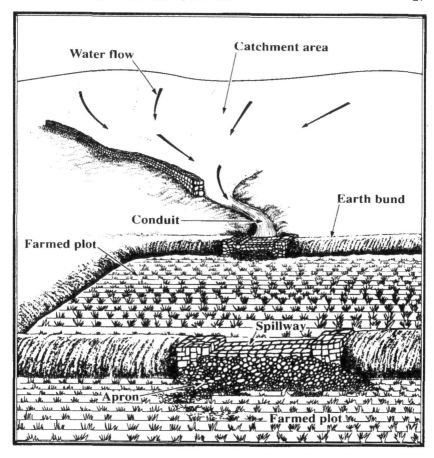

Figure 5   *The Israeli system of runoff farming, showing the catchment area with its low bund for diverting runoff water onto the field (or 'farmed plot'), and also the bunds which hold back water on the field. Spillways with stone facing are incorporated in these bunds so that any excess water can flow away without causing damage. (Source: Pacey and Cullis, 1986)*

Table 3 compares these two methods of rainwater harvesting with the traditional approach used in Turkana sorghum gardens.

PEOPLE AND AGENCIES

When development workers acquire an enthusiasm for a particular technique, there is a danger of it becoming a personal commitment being pushed ahead regardless of local reactions. While this kind of thing certainly occurred in Turkana, the more successful rainwater harvesting

Table 3   Contrasting techniques for rainwater harvesting

	Turkana method (site selection technique**)	Yemeni method (spate irrigation)	Israeli-style (Negev method) (runoff farming)
Source of water	runoff from natural catchments or flooding of seasonal rivers (Table 2)	floodwater in a wadi or seasonal river (Figure 4)	runoff water from catchments adjacent to fields (Figure 5)
Method of water control on the cultivated area	site is chosen for natural concentration of water	bunds spread water over land and slow down flow	bunds hold back water, ponding it on the land
Earthworks and other structures	none (except thorn barriers; ASAL 1990)	bunds, diversion structures (weirs) in river beds	bunds with spillways; conduits or banks on catchments
References	Morgan 1974; Pacey and Cullis, 1986, pp. 152–4	Thomas, 1982	Evenari et al., 1982 Critchley and Siegert 1991, pp. 75–86

** All traditional sorghum gardens in Turkana District depend to some extent on water flowing onto the site – crops are not solely rain-fed. The skill of the gardener lies in choosing a location where rainwater or floodwater harvesting is already taking place as a result of natural hydrological conditions.

techniques were developed by means of a process which was based on a dialogue between development workers and local people with their traditional techniques and immense, detailed knowledge of the Turkana environment. One purpose of this book is to show how this dialogue-based process worked by describing the interaction of different kinds of knowledge and experience within an evolving institutional framework. Two of the individuals first concerned were Paul Latham of the Salvation Army who suggested the relevance of the Israeli work in Turkana, and Francis Hillman, the volunteer worker recruited to adapt the Negev technique to local conditions. Francis did much valuable work during his two-year placement, and when he was leaving, the Salvation Army asked for another volunteer which is when Adrian Cullis was first involved. In Adrian's words:

As I had spent several months working at Avdat in the Israeli Negev and was interested in finding work on water harvesting, I was offered this posting by Voluntary Service Overseas. I arrived in Kenya in August 1979, and following a month's Kiswahili course, travelled north with Francis to Lokitaung. After an introductory month there, I was fortunate to have the opportunity to re-visit the Negev with Paulo Meyen, the project's assistant manager, to join a short course on runoff agriculture at the Avdat Research Station. Paulo was quick to recognize the potential of rainwater harvesting, but on returning to Kenya, we realised that soil and water conditions in Turkana were different from those in Israel, not least because our rainfall came in spring and summer, when evaporation was high, while theirs came in winter.

One aspect of the work done at Lokitaung in early 1980 was an effort to improve construction methods so that bunds would need less maintenance. Using terracing skills developed in Israel, attempts were made to construct larger bunds 1m high and stone-faced. In this way, it was hoped to trap adequate floodwater for maturation of a sorghum crop from a single flood, whilst at the same time reducing bund breaching. (Figure 6). Results were at first encouraging, but the rains in late March and April 1980 were heavy. In one period of only seven days, 175mm was recorded, including two storms of more than 50mm (Cullis, 1981). Many of the 'improved' bunds were breached and substantial amounts of impounded water were lost. It was later realised that the gardens which remained undamaged collected runoff from smaller catchments of between four and five times the cultivated area. Other gardens on the Negev model, had much larger catchments, and collected too much water, the spillways were under-designed, and bunds were overtopped by the floods. Despite the damage, it was possible to establish some good crops which produced yields of up to 1150kg/ha. Several families were therefore able to purchase livestock with surplus sorghum at a time when many other herders were totally dependent on food aid.

In March 1980, a European Community (EEC) delegation visited Lokitaung in the course of planning the rehabilitation programme described in the previous chapter. Travelling to Lokitaung from Lodwar, the delegation passed two roadside gardens which had harvested runoff during rain which fell in January. The gardens had mature crops when the delegation passed, and its members requested a meeting with project staff. The delegation became enthusiastic, and later proposed that water harvesting be incorporated into the rehabilitation programme. From this originated the massive TRP bund-building food-for-work programme despite advice that any developments should proceed cautiously.

Meanwhile, the Salvation Army needed a new source of funds to continue their water harvesting work, and in 1981 Oxfam approved a grant. It is interesting to note that the experience at Lorengippe in the 1960s was not taken into account and Oxfam sought to encourage the introduction of large spate irrigation systems. The rationale appears to have been that large-scale cropping 'could assist destitute Turkana to gain an adequate living from crop production' (Oxfam files, 1981). By contrast, the Lokitaung project originally had the more modest aim of *supplementing* the pastoral diet through improved sorghum production.

TRP AND RAINWATER HARVESTING

Following the visit to Lokitaung by the EEC delegation, the Turkana Rehabilitation Programme (TRP) became committed to the introduction of rainwater harvesting techniques for improved crop production and environmental rehabilitation. Expatriate volunteers were made responsible for co-ordinating both the relief programme and beginning the rehabilitation phase. The problems encountered have already been discussed in detail in the previous chapter. What can be added now is that during 1982, after a visit by an agronomist, Oxfam agreed to provide support for TRP's water harvesting programme in addition to working with the Salvation Army. Once again the emphasis was to be on Yemeni-style spate irrigation systems which would be used for improved crop production.

Great emphasis was placed on people being able to 'see for themselves' that water harvesting works, and it was proposed that Oxfam staff become directly involved in developing water harvesting sites for TRP. Thus Brian Hartley of Oxfam agreed to select suitable sites in north-west Turkana where Yemeni-style systems could be introduced. An Oxfam-recruited surveyor would then draw detailed plans. Attention was also given to the need to select groups to develop gardens more carefully, and it was proposed that the Diocesan Education Team should assist Oxfam to identify people to plant the selected sites. Whilst the importance of working with traditional social groups was recognized, it was suggested that TRP should retain control of demonstration sites to ensure effective maintenance and the introduction of improved cropping methods. Furthermore, despite the suggested early focus of the work in north-west Turkana, it was later agreed that Oxfam train a team of surveyors to develop Yemeni systems more widely in the district.

A volunteer team leader was appointed to co-ordinate the work, and fifteen Turkana site 'facilitators' were selected for training. These trainees were given basic instruction on site selection, surveying, design and construction over a period of several months, after which they returned to their home areas to begin work. As a result of the training, TRP made bund building the major effort of the food-for-work programme in 1983. In a period of ten months, an estimated 100 kilometres of bund were constructed at 120 sites in north Turkana (Finkel, 1984).

Enormous problems developed during execution of this programme, with expatriate volunteers and local site facilitators overwhelmed with routine work and unable to make progress and with people generally unconvinced of the value of what was being attempted. Meanwhile, in 1982 and 1984, two FAO consultants visited the area. Their reports argued that rainwater harvesting was the only means of raising agricultural production without recourse to full-scale irrigation (Powell, 1982) A variety of techniques which had not so far been emphasized in Turkana, such as trapezoidal bunds (Figure 6), were suggested. Estimates of 120 000ha of land suitable for water harvesting were given (Powell, 1982), and that 4250 families could be supported by one idealized scheme (Finkel, 1984). Among more useful comments, however, was one about the need to take account of social constraints and seek techniques compatible with the culture of nomadic pastoralists (Finkel, 1984).

Apart from Oxfam's attempt to extend the use of Yemeni-style spate irrigation in connection with the TRP programme, Oxfam continued to support the Salvation Army water harvesting project based at Lokitaung.

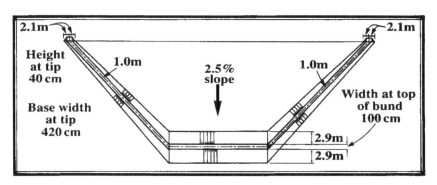

Figure 6   *Plan of a trapezoidal bund. This is drawn on the assumption that the land is sloping so that runoff water during a storm will flow from the right-hand side of the page to the left. The bund is designed to trap some of this water, flooding the plot enclosed to a maximum depth of 20cm. The maximum height of the bund is 60cm, and the other dimensions are given in metres. Although this was not adopted as a standard design, many gardens from 1984 onwards were made within bunds roughly trapezoidal in plan, varying with the terrain. The design was also modified by: reinforcing one or both tips of the bund with stonework to make a lateral spillway; digging a cut-off drain upslope to control excess flooding; levelling the garden enclosed within the bund to achieve an even depth of flooding. (Source: Finkel 1985)*

During 1982, this work was expanded in Kachoda and sites were also opened on the lake shore and at Kaikworr (60km west of Lokitaung). The project also introduced donkey-draught systems for cultivation and transport. Unfortunately, however, food relief support from CRS was discontinued in late 1982 and the project became dependent on TRP. Then the volunteer responsible for the Salvation Army work joined TRP. This move had Oxfam's full support on the basis that he would continue to support Lokitaung one day each week, in particular, to support Paulo Meyen, the Turkana project manager. However, things did not work well, and by 1984, the Salvation Army project was effectively limited to work with only one group of gardeners at Manalongoria (8km west of Lokitaung), where Francis Hillman had originally started in 1977.

At this point it needs to be noted that, despite Oxfam's emphasis on spate irrigation, little had been achieved by late 1983 in this direction. Most construction work and all successful sorghum gardens had been based on the Israeli technique of runoff farming or on the continued use of traditional Turkana methods. As a result, discussions were held with TRP, and it was concluded that a 'demonstration project' was needed in which different water harvesting methods – including runoff and spate irrigation systems – could be constructed under proper supervision.

The point here was that although there had been many experiments with rainwater harvesting in Africa, there had been little systematic work to show which techniques worked best in specific environments. It was not clear to what extent Middle Eastern experience, whether Yemeni or Israeli, could be used directly as a guide. In Turkana District, TRP had turned to water harvesting for its food-for-work programme. But because TRP was initially a famine relief agency and was still responsible for large-scale food distribution and its staff were overstretched, it could not easily experiment. What was needed was a separate project, specifically focused on experiments with water harvesting techniques and their social organization. The results of this could then help guide future work by TRP. The Turkana Water Harvesting and Draught Animal Demonstration Project was set up for this purpose as a two-year programme which was to begin on 1 January 1985.

The formal objectives of the project were as follows (Hartley, 1984):

a) to demonstrate water management systems, crop production, and range management systems applicable to the various conditions found in Turkana;
b) to establish work norms so that quantified food-for-work operations supported by the World Food Programme could be used by TRP;
c) to investigate the socio-economic consequences of introducing water management systems; to establish a dialogue with those who may claim usufructory rights; and to establish institutions to manage developed lands;
d) to introduce animal draught and transport systems applicable to the development of water management, crop production and forage production.

The aim was to demonstrate the full range of water management techniques. Particular emphasis was given to harnessing spate flows, with animal-draught power enabling larger systems to be built. With more reliable yields, it was thought that this would afford better opportunities for destitute populations to gain a livelihood and avoid long-term dependence on food-for-work (Hartley, 1984). The initiative was, however, complicated when the work was divided between a number of agencies. NORAD became responsible for the training element in the programme. Funds from the Netherlands government were allocated to make possible the appointment of a senior co-ordinator. Meanwhile Oxfam and TRP became responsible for establishing smaller demonstrations of spate irrigation and animal draught power, which it was agreed would be sited at Lokitaung. Coincidentally, the Intermediate Technology Development Group (ITDG) was seeking to expand its agriculture and water-related work and was looking for opportunities in Kenya. So as rainwater harvesting in Turkana entered on a new phase, ITDG became involved for the first time.

NEW BEGINNINGS

Adrian's assignment to the Salvation Army project had ended in late 1981, but at the time when discussions about a demonstration project were reaching a conclusion, he was visiting Kenya again, this time as an employee of ITDG. The aim was to identify projects which ITDG might support.

During this trip in March to May 1984, I visited Lokitaung. Discussions here prompted a visit to the Oxfam office in Nairobi, as it was clear the Salvation Army work had stalled. During the discussions, the Oxfam staff expressed interest in ITDG involvement, and I was eventually seconded to manage the work at Lokitaung. Returning in November 1984, it became rapidly apparent that local people were now conditioned to food-for-work. There were other difficulties, including the concern expressed by some TRP staff that the project would simply duplicate on-going work. More seriously, however, Oxfam had failed to alert the Salvation Army (at that time still receiving Oxfam funds) about the suggested collaboration in which they were to be involved. I therefore decided to explore possitle opportunities to support on-going work, rather than initiating new project work. Within the first few weeks, I spent a lot of time visiting both families living in Lokitaung and herders outside. It became apparent that many of the families from the more northerly areas who had been forced to move to Lokitaung as a result of *Lopiar*, had now returned to their traditional grazing grounds. Many had also succeeded in rebuilding viable herds and were now living more than 80km from the closest food distribution centre. Such were the suggested concentrations of herders and livestock, it was difficult to believe that a few years earlier, many of these families had been dependent on famine relief food.

Other familiarization visits were made to improved and traditional gardens, particularly along the lake shore, where it was possible in the early part of January 1985, following rain in November, to walk among mature stands of Turkana sorghum. In contrast, however, to the traditional gardens which had been planted and were being tended by their owners, many 'improved' gardens had been left unplanted. The people claimed that they were only maize gardens.

While it was good to meet friends and acquaintances again, I particularly wanted to spend time getting to know a small group. I therefore decided to renew contacts among the people of Manalongoria, to the west of Lokitaung, with whom I had previously worked. Accompanied on my first visit by Paulo Mayan, the Salvation Army project manager, we looked at gardens developed with help from TRP and met some of the leaders. Later, a traditional meat feast was organized for group members, at which it was hoped to initiate an informal discussion group. Thus it was possible to learn more about pastoralism, traditional agriculture, herding organization, and recent developments in the district.

The elders who took part in these discussions provided valuable insights into life in herding groups and famine camps, some of which will be presented and discussed in the next chapter. It would be untrue, however, to suggest that the elders always provided accurate, unbiased data. However, in time the group came to function very usefully as a local team of specialists who, unknown to themselves, challenged much of the accepted wisdom of the development experts in the district.

One point to emerge from these discussions was that once food-for-work was discontinued, all interest in improved sorghum gardens would evaporate. Increasing numbers of people were abandoning the famine camps as herds were rebuilt. Unfortunately, too, many of these families had become cynical about development and the possible benefits it could bring. It was therefore proposed to work on two fronts – firstly, investigations aimed at a better understanding of pastoralism, and secondly, support for the Salvation Army rainwater harvesting work in Kachoda. An earlier review of the Salvation Army's work suggested that work was at a standstill because nearby TRP garden improvement schemes offered food-for-work on a scale with which the Salvation Army could not compete. The challenge was therefore to attempt the reconstruction of a credible programme by identifying local needs which TRP had overlooked.

A detailed memorandum outlining these proposals was therefore drafted. I suggested that initially, three months would be devoted to learning from the experience of local people, in particular, traditional gardeners and those working on food-for-work schemes. The memorandum also identified areas needing to be better understood, including specific aspects of traditional gardening, such as choice of sites for gardens, and land tenure rights. I also wanted to identify appropriate end uses of seasonal runoff water which the Turkana themselves would be likely to utilize to optimal effect within their pastoral system. For example, some remarked that they had tried to impress upon TRP surveyors the importance of levelling gardens so as to spread flood water more evenly over sorghum gardens. This suggestion was ignored. Finally, I wanted to gain a better understanding of appropriate institutions which would enable local herders to establish and maintain control over improved lands.

3. THE PASTORAL ECONOMY AND TURKANA INSTITUTIONS

LAND AND FOOD

A central function of Turkana families is to develop and refine strategies which ensure survival of livestock during the long, often harsh, dry season. Such strategies include keeping mixed herds of several species which are adapted to different ecological habitats, and moving herds regularly from one area to another according to which lands provide the best grazing at different times of year.

Other strategies which contribute to the survival of both people and animals include splitting a herd at certain seasons and leaving some people and livestock in a home area (known as the *ere*); and using a variety of food sources to supplement the livelihood gained from animals. In the latter context, for example, grain is consumed in large quantities at the end of the dry season when milk and wild fruits are in short supply. Animals are also bled and slaughtered in the dry season as blood and meat then replaces milk as the staple food. The main wet season food (Figure 7) is buttermilk, while fruits and berries are collected by the women and girls in season and are brought home to be prepared and shared among the family (Watson, 1988). In all, 47 species of wild plants are used by the Turkana to a greater or less extent (Morgan, 1980).

Herd-boys enjoy hunting and can seldom resist giving chase to rabbits and other small game which they disturb on the grazing lands. Older youths and adult men also engage in more serious hunting but in recent years have substantially depleted game stocks in the district. Concentrations of zebra, oryx, eland and other antelopes can now only be found in the more northerly border regions. Herders say that in times past, small groups of Turkana existed mainly by hunting as they had too few animals on which to survive. Hunting groups frequently lived in relative isolation from herding groups as the former followed game migrations.

For pastoral groups living on the lakeside, fishing has long been a means of supplementing other foods. The traditional technique was to plunge wicker baskets down onto fish feeding in the shallows. As mentioned previously, the use of boats and nets to catch fish for sale began only in 1961. Much of the cash earned this way has been invested in livestock, thereby reinforcing the pastoral tradition (Broch-due, 1983). Many fishing

	Jan-Mar	Apr-May	Oct-Dec
Meat	******************* (slaughtered/ dead animals)	* * * * * * * * * * (ritual slaughter)	***************
Milk		********************************** * * * * ****** * *	
Blood	************	* * * * * * *	**************************
Sorghum	******************		************
Fruits		*************************** * * *	

Jan Feb Mar Apr May June July Aug Sept Oct Nov Dec

Rainy season   Short season

Figure 7   *Seasonality of the more important foods consumed by Turkana pastoralists. (Blood is obtained by careful bleeding of live animals.) (Source: Watson, 1988)*

families are also actively engaged in seasonal sorghum cultivation, with the result that household economies along the lakeside are more diversified than elsewhere.

Cash plays a relatively small part in the pastoral economy, as all herder/ herder transactions are carried out using livestock as the medium of exchange. Somali traders who travel into the more isolated and remote parts of the area (often on foot), commonly trade with livestock, which are then driven back to the settlements and sold. Following the 1980 famine, however, large numbers of herders were forced into the settlements. Since then, and despite the return of many families to herding, cash appears to have acquired more importance.

The chief advantage of the strategy of keeping a variety of livestock species is that some animals (such as cattle and sheep) are grazers living mainly on grass and herbs, while others (camels, goats, donkeys) are browsers, feeding from bushes and shrubs. The extensive grasslands to the north of Lokitaung favour large cattle herds, whilst significant numbers of camels and goats are kept in the drier plains to the south. Different species also vary in their susceptibility to disease. Keeping several types of animal thus reduces the risk of an epidemic wiping out all of a herder's stock.

Another reason for keeping a variety of species is that the quality of their milk and the length of lactation varies. Careful management of a mixed herd can ensure that some milk is available almost all the year round (Swift, 1981). Herders building up their flocks generally start with small stock and then diversify by exchanging goats or sheep for larger animal types according to long-established exchange rates, such as six goats for a male donkey, or eleven for an ox or bull. In turn, male animals can be exchanged for young breeding females. During the 1980s, when many families were rebuilding herds after *Lopiar*, goats predominated in many herds and cattle were few.

Whatever the advantages of keeping a variety of livestock types, their divergent browsing or grazing requirements become increasingly incompatible during the dry season. Turkana *zebu* cattle are better browsers than many other breeds, but still require access to grasslands which are to be found mainly in the northern border regions and in the mountains. It is therefore common practice for families owning large numbers of cattle to have a highly mobile camp and to move their cattle widely in search of good grazing, while their browsing stock (camels and goats) is moved less frequently and remains closer to the wet season grazing lands.

LAND TENURE AND 'SECTIONS'

The seasonal movement of their animals is the most important of the drought-avoiding strategies used by Turkana herders. However, to understand these migrations it is necessary to appreciate something of how local land tenure systems work. From the point of view of government, Turkana grazing land is classified as County Council Freehold. Within settlements, a plot can be registered under a private name with the County Council office in Lodwar, for which a small annual payment is levied. Registered land is effectively privately owned, as title deeds are held. Apart from one holding-ground established in 1987, no major grazing areas have been enclosed although land has been lost to imigration schemes. Rangelands continue, therefore, to be used in accordance with the traditional Turkana land tenure system.

The Turkana appear to deny the existence of any system when they say 'Turkana can go anywhere'. They emphasize the ability of herders to move freely about the District with their stock (Gulliver, 1951). But investigation reveals a complex system of rights in land. These include the territorial 'section' (*ekitela*) within which grazing and water resources are controlled; the *ekwar*, which is land alongside the bed of a seasonal river on which there are trees whose use is controlled by the owner (Barrow, 1988); the *ere* or 'home area', where the graves of ancestors and sometimes gardens are located; and the larger sorghum gardens (*emanikor*).

The *ere*, which is where families return in the wet season, assumed particular importance in the work to be described later because it appeared as the most appropriate land-based unit for extension work. Families have long-standing connections with particular *ere* and *ekwar* areas, and also with gardens. Rights to use different areas are complex. With gardens, much depends on whether rights were acquired by clearing the land, or by inheritance. With riverine land (*ekwar*) herders have exclusive rights to the

Sheep and goats returning to a Turkana homestead in the evening. In the centre are the day and night houses, which every married woman has.

Milking is done in the mornings and evenings by the women and girls. The kid goats are allowed to suckle at the same time as the goat is milked into a wooden container.

Making a wooden milk container, using an adze to carve the piece of wood. Most containers are made from wood, with goatskin lids and carrying straps, although gourds are used for churning milk.

Learning how to use the line level. The level is used for surveying a garden site, setting out the bund design, and levelling the garden surface.

Aligning the sticks which show where the bund will be built. Surveyors carry measuring sticks with notches marked at certain heights to facilitate the demarcation of the garden site.

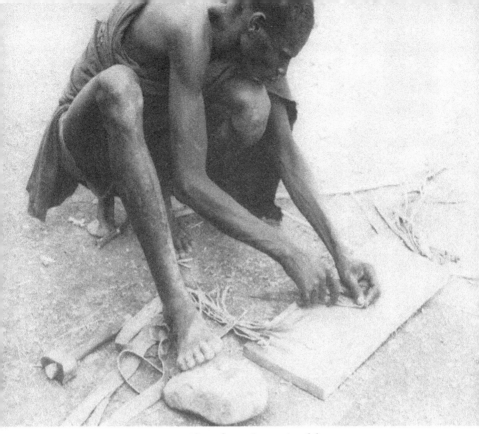

Manufacturing animal harnesses.

An early stage in building a bund. The garden has been surveyed and pegged out; the sticks mark the slopes and crest of the bund.

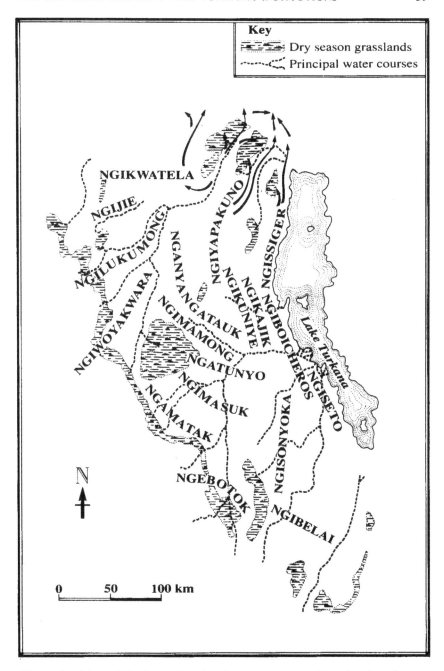

Figure 8   *Map of Turkana District showing names of 'sections' and territories they occupy. Arrows indicate dry season movements of herds in three northern sections. (Source: Gulliver 1951)*

pods and leaves of groups of trees, in particular those of *Acacia tortilis*. This makes it possible for herders to manage scarce resources in a rational way. Barrow (1988) found that 77 per cent of *ekwar* along the Turkwel river were inherited, and that roughly 60 per cent had been held by the family for more than two generations. Further north, however, there are no *ekwar* as herders are more mobile.

What characterizes all these concepts of land tenure – and indeed all other social institutions including the homestead (*awi*) and the grazing or herding group (*adakar*) – is their flexibility and their variation in space and time. For example, the elders of the sections or *ekitela* in the centre of the District are in general more involved in the control of grazing and water resources than in those sections at the periphery, for the simple reason that in the drier central areas where natural resources are scarcer, there is more need for some form of grazing regulation.

Similarly, the section is not fixed in time. As the need for it alters, for example with the deterioration of security in the northern areas, so the importance of its leadership and its central organization increases. It is this flexibility, coupled with an apparently informal leadership, which has led to the widely-held view that the Turkana are completely individualistic (Gulliver, 1951). However, a range of social institutions do exist which give Turkana society a structural coherence, with reducing flexibility and adaptation to local circumstances.

The division of grazing land among sections is particularly important for seasonal stock movements. Of the nineteen sections in Turkana District recorded by Gulliver (1951), three are in Lokitaung Division. Their names are Ngikwatela, Ngisiger, and Ngiyapakuno, and the approximate extent of the territory they occupy is indicated in Figure 8. This map shows that herders in each section have access to specific areas of dry-season grazing land, often in mountains or other uplands. Whilst there is some movement between sections, it seems clear that most herders prefer to stay on the land associated with their section. Movements of livestock are far from random, and herders leave the territory of one section for another only in times of acute stress (Tornay, 1979).

Because of higher rainfall and lower temperatures, mountain pastures offer good quality grazing long after the grasslands on the plains have dried up. Movements into mountain ranges are controlled, for it is recognized that such pastures are a reserve for long dry seasons. The management of the mountain pastures is therefore a delicate matter of balance between short-term (milk production) and long-term (drought reserve) grazing objectives. Before herders move into mountain pastures, *adakar* leaders meet together to decide which parts of the mountain should be opened and at what time. This form of range management has been poorly understood by administrators and development workers alike, to the detriment of development initiatives in the District.

Sometimes herds move across into Sudan, as for example, the movements of three sections represented on the map by arrows (Figure 8). The arrows indicate how animals are grazed increasingly further north as the dry season progresses. Thus in 1984, a year of low and erratic rainfall, herders of the Ngikwatela section moved into southern Sudan, close to the Ethiopian border and the Merille people. They camped in large groups of 40 to 50 households (*arigan*), each protected by a stockade of thorns. At dusk, the gateways were closed by pulling large thorn bushes into the entrance, and armed guards settled into shallow dugouts ringing the inside of the stockade. In the mornings, livestock were kept within the camp until a group of heavily armed warriors returned from a scouting expedition around the camp.

Further evidence of this sort of problem was revealed by an interview with a herder named Edapal Mariao in September 1987:

Following the wet season in the Kotome valley, we began to move north with other herders. When the first *adakar* (herding group) reached the northern end of the valley, their scouts reported a large number of Toposa raiders (from Sudan) were encamped to the west of the Lokwanamoru range. We therefore decided to halt our northerly migration. Together with other herders, I moved my cattle into the Lorioneton mountains to the east, well away from the raiders. The large camps of camels and small stock remained in the valley, although ready to move at a moment's notice. In the following two weeks, our men attacked the raiders several times in an attempt to dislodge them. Eventually, we gained control of the water hole where the raiders had been camping, and as we had already established defensive positions around all the other water points in the area, the Toposa were forced to leave the area. Once the threat of raiding had passed, I brought my cattle back into the valley, where they remained as long as possible before moving further north. Meanwhile, I visited my brother – the Chief – for food and cash gifts. Our family and herds were among the last to cross the pass at the head of the valley to reach the plains (of southern Sudan).

In northern Turkana, along the foot of the Lokwanamoru and some other mountains, there are perennial springs which herders can use for watering cattle. However, there are several areas where the grazing is difficult to exploit due to lack of water, and there are superb grasslands which can only be used during the wet season as there are so few places where hand-dug wells can be made. In contrast, herders along the lake have no problem watering their animals. In dry seasons, when fresh water is no longer available in the seasonal rivers, both human and animal populations use the lake water, even though it has a high fluoride content.

In the wet season, most herders move onto the plains where they reunite with other family members from whom they separated in the dry season.

Following good rains, the plains offer excellent grazing. Water is freely available, and movement is much easier than in the mountains. The beginning of the rains can be a difficult time, however. If animals have been weakened by a hard dry season, they suffer shock and can develop pleural pneumonia following heavy rain and the associated sudden drop in temperature. When Edapal was interviewed, he told us about the beginning of the wet season in 1987:

I could hear the distant rumbles of thunder over Lorionetom and Lokwanamoru (mountains) far to the south. Later, just before daybreak, I got up and called to Alim (my first wife) to tell her to re-tie the hides covering her *akai* as the storm was moving in. The winds began to pick up as she worked. Within minutes of the first drops of rain the camp was awash with water as the storm lashed down. The rain continued until mid-morning, after which the sky began to brighten in the east.

I crawled out of the *akai* and could see other herders beginning to move around in the camp. It was cold and the ground was sodden. As I walked towards the goat kraal, so much mud stuck to my tyre sandals, I took them off. The goats were standing with their heads down and backs arched. I lifted back the thorn bush (which was pulled into the gateway for the night) and could see a number of dead goats which had fallen victim to the sudden cold. There were twelve dead. As the sun was now breaking through, I opened the gateway and let the goats into the main thorn enclosure. . . .

ORGANIZATION OF LIVESTOCK MOVEMENTS

Section grazing is managed by elders and it is known, for example, that herders are prevented from moving their animals into dry season grazing too early, to ensure an adequate reserve. Hogg (1982) makes the point that in times of stress, grazing rights 'crystallize' into more formal patterns of ownership. Access may be denied to members of other sections in order to protect grazing for members of the home section. However, it is also true that in times of crisis, herders support each other and share depleted grazing. At such times, herders who cross a section boundary often present their hosts with livestock (McCabe, 1985).

Each section in Lokitaung Division has its diviner, who is the magico-religious leader. These powerful figures (occasionally women) are known to each member of the section and are able to exert significant influence over herding migrations, raids and ritual activities. After the long rains in 1986, for example, at the start of which many livestock died, a diviner told one sub-group of the Nkigwatela to move into the southern part of the Kotome valley, and perform a particular ritual there. As a direct result, all the households moved south, resting only one night between moves (which is unusual for the Turkana).

Within the boundaries of the section, herders co-operate in the small grazing groups known as *adakar*. These are clusters of homesteads which come together to share herding duties, especially in the dry season when the work burden is higher. At this time, livestock are frequently divided into two separate herds, as already explained, with the cattle which need grass travelling more widely than browsing animals. For example, Watson (1988) mentions one family whose household moved only five times in fifteen months, whilst their cattle moved camp eight times. Most herds move more often than this, but much depends on local conditions, the mix of animals kept, and individual judgements by herders. Watson records one man who was an enthusiast for frequent moves and exasperated other members of his household by moving twenty five times in fifteen months. McCabe and Fry (1986) monitored herd movements in Kakuma Division, and their data revealed that individual households sometimes moved differently from the overall pattern which the section should have followed. A number of factors which distorted 'ideal behaviour' (wet/dry season movements) were identified as raiding, disease, and the availability of food aid. In Lokitaung Division also, many herders had abandoned their usual nomadic movements at the time referred to (1982-3), sometimes because of loss of livestock. However, even herders with good numbers of animals were drawn out of their traditional migrations toward settlements where food aid was available. By 1985, however, many herders of the Ngikwatela section are known to have resumed more traditional migrations, and large herding groups (*adakar*) were found again.

In the northerly border areas, it appears that considerably more raiding takes place in the wet season than at other times, and security is therefore an important consideration. That is one reason why most pastoralists move their herds south for the wet season. Large concentrations of people and many thousands of animals gather then in the Kotome valley (which is more easily defended), a fact now appreciated by the Ministry of Livestock as an opportunity to reduce the logistical difficulties of vaccination campaigns.

FAMILY LIFE

The family unit lives in a mobile camp or *awi*, which has simple, hemispherical huts made of strong interlaced branches. Smaller branches and leafy material complete the roofs of day huts, whose function is mainly to provide shade. By contrast, when the night huts are in use, they are covered with hides. In larger camps, each wife has her own thorn enclosure into which the animals for which she has milking rights are herded each night. The men sleep outside in a simple enclosure of thorns which acts as a windbreak. When the household moves to new pastures, the huts and enclosures are abandoned, though it is common for the huts to be used by other herders in the area as a source of building materials or firewood.

The male head of the household is ultimately responsible for all herding decisions, although he will normally consult others in the home. Family members will be made responsible for different groups of animals, separated not only according to livestock types, but also between different groups of the same species. Herding cattle is one of the more arduous tasks, particularly in the dry season when cattle camps are moved into the mountain pastures. Then the animals often have to trek long distances between grazing and water. Despite the extra work involved, most young men are anxious to join cattle camps, possibly because there is considerably more status attached to cattle than other livestock. All men and youths have a 'dance-ox' (or a goat if the family has no cattle) about which they sing on occasion. When visiting camps, one can sometimes see a 'dance-ox' being smeared with dung by its owner in the early morning as it is paraded around with the owner singing his 'ox-song'. Sheep and goats are divided into four flocks. The very young animals are kept together in the shade, usually tied up. Slightly bigger animals may be herded by the younger children near to home, where the women can keep an eye on them. These animals graze, but water is brought to them. When young animals are big enough to travel easily to water, they join the breeding herd, which is usually looked after by boys. The milking herd may form yet another subdivision.

Livestock watering in the dry season is the combined work of both men and women, with the men supervising the herds at water-holes, while the women lift water from the hand-dug wells in seasonal river beds. Women deepen the wells as the dry season continues, and then work together to lift water up a 'human ladder'. Sometimes four women can be seen working together in this way, though more often two suffice. Animals drink from troughs made from tree trunks, but large aluminium cooking pots have recently come into use for this purpose.

During the long rains, herding demands are significantly reduced and this is the time when various ceremonies take place, including many marriages. Women and girls also take the opportunity to prepare skins and hides for clothing, sandals, and shelter. In good years, when herds are producing a surplus of milk, the women prepare a form of dried milk powder called *edodo*. This can be stored for many months without deterioration, but most is reconstituted with fresh milk and mixed with dried vegetables or meat later in the same wet season. It seems that rather than storing large amounts of food which would be difficult to transport, families build up their own body reserves. They gain weight in the wet season, expecting to lose it in the dry months, which is a well-documented practice elsewhere in Africa (Pacey and Payne 1983, p.28).

Milk processing is the focus of an interesting and highly developed web of interacting techniques or 'technology complex' involving not only methods of making dried milk or buttermilk, but also the containers used for handling and storage of milk. Some of these are carved in wood by the women using special iron tools (see photo section). Other vessels are made from gourds and are used for water-carrying as well as for milk.

Ownership of livestock is based on a set of rights and obligations, without any concept of an animal being somebody's exclusive property in the western sense. Each family member has rights to a number of animals in the family herd, but these animals are shared with other members, and may well vary over time. For example, a man may 'own' twenty cattle, but the milking rights are divided amongst his two wives and their children. He may be obliged to part with animals so as to support his brother in paying a bride-price, or a friend may request the loan of a male animal for breeding. It would be difficult for a herd owner to refuse to acknowledge these obligations, for to do so would put him outside the social system. Decisions concerning the use of the animals are not solely made by senior men (even though they may suggest that such is the case) but other family members and other herders are involved.

A family's animals are divided amongst the wives, and pairs of day and night huts identify separate economic units within the homestead. A wife together with her daughters will milk the animals (usually twice a day), process the milk, and allocate it to members of the family. Children are also allocated milking rights to particular animals, which are consolidated as the child grows up. In many cases, the animals will form the nucleus of a herd, on which the children will one day depend. Other non-milking animals may also be allocated to children, especially boys, who at puberty will seek to establish rights to a dance-ox. Gulliver (1951) describes how a youth's claim to the ox can be strengthened by putting a collar and a bell around its neck.

Herders have pointed out that rights to meat are different from rights to milk and blood and it is the men who decide how frequently animals may be slaughtered. This said, however, it appears that men may only slaughter their own animals and cannot remove livestock from their wives' kraals without prior consultation. Furthermore, in times of hardship, wives can put considerable pressure on their husbands to slaughter animals.

SOCIAL NETWORKS

The Turkana have distinct social networks through which each individual is related to others by giving and receiving relationships (usually involving livestock, livestock products or sorghum). The network includes relatives, in-laws, and 'stock friends'. Marriage and the exchanges of livestock which follow are especially important in reinforcing links. Networks are of particular importance in a crisis, when the existence of an associate who lives outside the area struck by livestock disease or a raid can mean the difference between pastoral survival and destitution.

One aspect of social networks is the clan structure among the Turkana which has been discussed by social anthropologists. Gulliver (1955) identified a total of twenty eight exogamous clans or *emacar*. Each clan or extended family has its own set of taboos and restrictions which include styles of dress and hair, together with specific ritual practices observed during menstruation, weddings, pregnancy, burial, and other significant stages in life. Herders report that they are able to claim hospitality from other members of their clans when travelling far from home (both within and beyond the borders of the section). Similarly, it is thought that clan links between sections may serve to ease tension and enable rights of passage to be negotiated during times of drought. Clan links may also be activated by herders when members of a section have lost large numbers of animals and are therefore unable to offer support to one another.

Clans and socal networks sometimes cross lines of ethnic origin (Schlee, 1985; Sobania, 1975). Studies carried out among the Donyiro and Merille peoples in Ethiopia have shown that their social links include Turkana 'stock friends'. Intermarriage also occurs. Among project staff in Lokitaung, for example, Namongo the blacksmith had a Donyiro mother, though he is Turkana, whilst Ayanae's father was Merille and her maternal grandfather was Karamojong. Tornay (1979, p.168) suggests that linguistic differences are not a barrier to positive contacts and exchanges because pastoralists frequently know more than one language.

All adult men belong to a 'generation set', which is subdivided into age-sets, members of which become closely 'bonded'. For example, members may engage in raiding together, or assist one of their number raise a bride price. Studies carried out among other ethnic groups in the region suggest that age-sets establish a social hierarchy which enables the elders to exercise social control over warriors (Tornay, 1979). In this way, elders are able to prevent raids from escalating into full-scale tribal warfare which would theaten social stability. Traditionally, raiding and warfare evoked a degree of temporary animosity, but clan and network links remained. Through them, peace treaties were made, and spears broken and buried. More recently, following introduction of automatic weapons, the social etiquette of raiding has been eroded, and many more lives are lost, as mentioned in Chapter 2.

Through understanding the social and economic functions of networks, we can begin to see why herders attach such significance to extending them, so that animals are more often invested in strengthening a herder's social network rather than being sold for cash. Sorghum is also used in this way. Neither is it only men who are involved in livestock transactions of this sort. In Turkana, visitors may be asked to transport goats which are gifts from women to women.

Relationships established by marriage are particularly strong in Turkana and in-laws are frequently found herding together. Whilst marrying off a daughter is a means of widening a family's network, taking a wife requires herd-owners to draw on already existing networks for livestock to meet the bride-price. In recent years, following recovery of the pastoral system from the *Lopiar* famine, bride-prices have risen steeply and commonly amount to between 20 and 50 large stock, and 100 to 150 sheep or goats. To gather such large numbers of animals and at the same time to maintain viable herds, necessitates the full support of a herder's network.

TRADITIONAL AGRICULTURE

Whilst the most important Turkana institutions are those related to herding, other institutions are connected with the production of grain and trade in sorghum. There is ample evidence that sorghum cultivation has been a significant activity in Turkana for a very long time (Gulliver, 1951; Henrikson, 1974).

The most productive sorghum growing areas tend to be along the major river systems and drainage channels. The Kerio and Turkwel river valleys, in particular, produce large amounts of sorghum following the annual floods, and herders travel to both to obtain grain by trading. In years of ample river flooding, traditional plots on the Turkwel river produce bigger harvests than the Katilu irrigation scheme (Broch-due, 1983). Much of the cultivation is carried out by people of the Ngebotok section (Figure 8) in the upper reaches of the Turkwel where rainfall is higher and the river may flow for half the year. Here it is often possible for people to grow enough grain for a year-round supply.

In the north of the District the pattern is similar with significant areas devoted to sorghum production along the Tarach river in Kakuma Division (Figure 2). Gulliver (1951) also identified a large area of sorghum in the northern part of the Kachoda valley (Figure 16), but this has been abandoned for many years. It is possible that border raiding between the Turkana and the Merille is the reason why these gardens are no longer used. In the southern part of the Kachoda valley, the Ngisiger and Nigyapakuno sections have made little effort to establish sorghum gardens, although it is understood that a few families have planted on an irregular basis. The only sustained attempts to cultivate gardens appear to be at Manalongoria, and two other places. These gardeners are known to be descendents of Turkwel valley cultivators.

Apart from this, gardening in Lokitaung Division has latterly been confined to the lake shore. This area is extremely arid, but rainwater drainage to the lake is impeded in places by a ridge of sand-dunes (Hillman, 1980). Deep, fertile silts have built up, substantially increasing the water-retaining capacity of soils. Along the lake shore south of Todenyang and around Loarengak, there are at least 200 sorghum gardens. Cultivators say that some were first made long ago by the Merille. Others were begun by Turkana from the Ngisiger section and the most recent are the work of destitute groups who were moved to the lake in order to take up fishing after famines in the 1960s.

Cultivators along the lake have provided detailed information on traditional agricultural practice, some of which has already been quoted (Table 2). Regular discussion groups were held by staff of the Demonstration Project over a period of several months in 1984–5. Gardeners stressed that inadequate rainfall was the main constraint on crop production, and emphasized the importance of selecting sites where runoff water concentrates. This makes it clear that they were consciously practising a form of rainwater harvesting and were not choosing garden sites randomly in ignorance of moisture conditions.

It was also learned that customary planting rights are established in a variety of different ways, according to how and when the garden was obtained. At the simplest level, herders can establish rights literally by 'seeing' and 'choosing'. It appears, however, that these rights are frequently challenged and the importance of clearing in preparation for planting was stressed. As areas inundated by flood water are frequently invaded by woody scrub, clearing is demanding work, which is commonly shared by both men and women. Customary rights associated with clearing are seldom challenged. Finally, full planting rights are established as a result of planting seed and after this stage the site is referred to as an *amana* (garden).

Established planting rights are handed down from generation to generation in much the same way that livestock are allocated. Each wife has her own garden for which she alone has rights to plant, harvest and dispose of the crop. When she dies these rights are handed on through her eldest son, to his first wife. Widows retain their rights until they die, when planting rights are inherited by the eldest son. In-laws, relatives and stock associates can request the 'right' to plant a small portion of a garden during a particular season. Such requests are apparently uncommon, but when made are difficult to refuse. Planting rights are therefore not to be confused with exclusive rights of ownership. Cultivators tell of attempts at compromise, particularly where good land is more freely available, and claimants are encouraged to clear an extra plot adjacent to the garden. In this way a woman can avoid subdividing her garden, which would involve separating plots with footpaths.

The staple crop throughout the district is a variety of *Sorghum bicolor* known as Turkana sorghum or *ngimomwa* (Gulliver, 1951). As a result of careful seed selection over many years, Turkana cultivators themselves developed this variety which matures remarkably quickly in 62 days. Other sorghum varieties found within the region are slower maturing, such as *Donyiro* sorghum (80–90 days) and *Merille* sorghum (90–100 days). Both have been developed in areas where aridity is not such a constraint on production and greater emphasis has been given to selecting higher-yielding varieties. Small amounts of other crops are cultivated by Turkana gardeners, including cowpeas, green gram, Ethiopian maize, water melons, gourds and occasionally millet. These crops tend to be planted separately from sorghum in small plots at the edges of the garden, rather than inter-planted between the rows of sorghum.

WOMEN AND SORGHUM CULTURE

Each wife in a family is responsible for storing her own seed between seasons, although seeds are shared in times of shortage. Women prepare small bags made from the skins of kid-goats or jackals, and these are tied securely to the frame of the night hut. Several women report that they try to prevent pest damage by dusting the seed with either fine red dust or ash, but it is not clear how widespread this practice is. The red dust is probably *emunyen* (also used by the Ngisiger for personal adornment), which is prepared by drying and crushing a form of red clay. The ash used for seed treatment is made by burning the wood of particular trees, including *Acacia tortilis*. Lakeside sorghum producers point out that, whilst they keep their own seed, they continue to trade for new seed with gardeners further south. One lakeside gardener, Apim, lives with her husband Eleyo close to the shore, and told project staff:

My husband herds some stock and our two eldest sons have a fishing boat. Following rain (in April 1987), our garden at Natieng (north of Loarengak) was flooded, and as the water receded, we began to plant in the exposed soil. Each day I put a handful of seeds into a tin partially filled with water, and left them to soak overnight. The next day, I took the seeds to the garden and planted them, using an old *panga* (machete) in place of a digging stick, and putting six or eight seeds in each hole. Working slowly in this way, my daughter and I completed the planting over a five-week period, by which time the first sorghum was well grown. We also planted cowpeas, which I was given by a friend. We had to spend much time weeding the garden to control the knot grass which grows there and returns year after year.

Knot grass is a major problem on the better-watered sites and in places almost chokes the sorghum before weeding is completed. Despite this, sorghum recovers quickly when weeds are cleared and when well established, suppresses further weed growth. There are also bird and insect pests to be controlled. Raised platforms made of poles are erected in the fields and from these, children shout at the birds, rattle tin-cans filled with stones and throw mud pellets at the more persistent flocks. To control insects, green plant material, donkey dung and dry twigs are arranged in small heaps along one side of the garden. These are fired in the evening when there is just sufficient breeze to carry the smoke into the crop. Insects may then be seen flying away. Repetition of the smoke treatment helps prevent heavy insect infestation building up.

Fencing gardens with thorns to keep livestock out is usual in the Kachoda valley and on the Kerio River (Morgan, 1974), but is not practised on the lake shore probably because suitable woody material is less freely available. As noted in Chapter 2, some authorities suggest that thorn barriers can be dense enough at ground surface level to play some role in water spreading (ASAL, 1990).

Prior to harvest, family members begin to eat the grain before it is fully ripe and they also eat the sweet stems. Cultivators tell how in good years, the first ripe heads are taken as a gift to a respected elder and other early-ripened heads are selected for seed. Both men and women take part in seed selection. Gardeners are able to identify more than twenty strains of local sorghum, each with slightly different characteristics regarding susceptibility to drought, head size, rate of maturing, tolerance of water-logging, and taste. Selected heads are threshed separately from the main harvest.

Much of the crop is stacked in the sun to dry and is then threshed by the women on a prepared threshing floor. When the threshing is complete, the grain is scooped up into a wooden bowl and slowly emptied out again, letting the wind carry away the husks and chaff. This winnowing operation is carried out several times and the grain is then stored in large goat-skin bags. Casual labourers are occasionally employed to assist in threshing and winnowing the grain. Daily rates of pay appear to vary enormously, but the daily ration of 1.5kg of grain introduced in the World Food Programme food-for-work schemes has become fairly widely adopted since 1981.

There are no reliable data for yield from traditional gardens in Lokitaung Division. However, one estimate is available from the Kerio River area, where a 5m square plot was thought one year to have produced the equivalent of 1280kg/ha (Morgan, 1974). Gardeners on the lake shore report plot yields of between two and three goat-skin bags plus 10–20kg seed in a good season. If an average estimate for the weight of a full bag is taken as 50kg and an allowance is made for sorghum eaten raw in the field, this amounts to a yield of 750–800kg/ha on roughly 0.2ha plots. Yields equalling those estimated for the Kerio valley, (around 1300kg/ha), however, have been recorded on improved plots. By contrast, on irrigated land in Turkana, yields in excess of 3000kg/ha are 'expected', although in 1982, the average yield actually achieved was 1235kg/ha (Broch-due, 1983). It must be remembered that, in addition to the grain crop, some sorghum also provides a large quantity of sweet stems which are eaten by family members.

People remember especially good years by name, including 1958 (*Ngisine*) and 1974 (*Akwadodo*). Bad years in which there is little or not harvest are also remembered, such as 1970 and 1972 (*Kimududu* and *Kibekebek*). These were seasons when sorghum crops failed entirely. In some such years, people recognize that the rains are not likely to be sufficient, and continue their dry season migrations without planting at all. This is not as serious as it sounds in that the grain produced in a sorghum garden is not a major part of the subsistence of a family. Rather, it is invested in ways which will make the family or group more secure in a bad year, by being exchanged for livestock so that there is a bigger herd to live off in the next dry period, or by being given to 'stock friends' to strengthen the family's links with its social network.

TRADE IN SORGHUM AND RESPONSE TO DROUGHT

The northern Turkana regularly trade with the Merille people who plant their crops along the Omo River. This flows south from the Ethiopian highlands into Lake Turkana. Flooding in the lower reaches occurs in August and September after the long rains in the highlands, and it is not until the floods subside that the Merille begin to plant. Their sorghum takes longer to mature than the Turkana variety and it is therefore only in January or February that the grain is harvested as compared with July or August for the Turkana crop. The Merille harvest thus coincides with the long dry season in Turkana, which is when access to cereals is most crucial.

In order to discover more about trade in sorghum with Merille and Donyiro peoples, and among the Turkana themselves, project staff interviewed 48 households in the Kachoda valley about their trading activities during the previous ten years. Of these, 40 said that they had exchanged livestock or goods with the Merille during one or more dry seasons. Householders were also asked about the items most commonly used for trade and the approximate exchange rates. We found that goats were most frequently exchanged, at the rate of one goat for 40–50kg sorghum. Skins were also regularly traded for 10–20kg grain each. People would also sometimes barter milk containers made from gourds, obtaining as much grain as the container would hold. These are all examples of purchases of sorghum, but people also sometimes parted with sorghum to obtain sheets of cloth (40–60kg per sheet), sandals (30kg per pair) or tea (2kg per 100g packet).

Some insight into how trade is organized was obtained when travelling in the far north of Turkana in November 1985, encountering sixteen donkeys laden with sorghum. The caravan was accompanied by twelve adults, including seven women and five men (three of whom were armed). In the discussions which followed, Adrian learned that some women in the *adakar* to which the group belonged had complained that their children were hungry. The leaders met together under a tree and decided to send a party north to visit the Donyiro people. The leaders then selected young men to travel with the party and families who wanted sorghum selected a goat and provided a donkey for transporting the grain.

When the group arrived at Kibish, they dispersed to visit Donyiro friends with whom they had traded in the past. It appears that some Turkana have long-standing trading relationships with particular Donyiro households, and whenever possible renew this contact. Gulliver (1951) suggests that some Turkana have relatives among the Donyiro with whom they combine trade and family visits. This party told me that some Turkana visit the Donyiro as friends and are not always expected to bring livestock or goods to trade. Some families are expected to take goats only on alternate visits. While this may appear to offer very attractive terms of trade, Turkana families doubtlessly reciprocate in other ways.

Despite the difficulties they face during the dry season, herders are sanguine about their lot. They point out that living without rain has become a way of life, something they are good at, and therefore not something to worry about. In Lokitaung during late 1986, the short rains of October and November failed and there was a seven-month period without significant rain (worse for the area than implied by Table 1). Local pasture and browse deteriorated rapidly and many of the local water sources failed. Watering problems faced by herders were compounded when townsfolk from Lokitaung began using rural water points because the town supply had broken down. The additional demand resulted in herders waiting up to four and five hours before starting to water their herds. Time spent at the wells was time taken from grazing and animals lost condition rapidly. Herders said that some animals had 'lost their sense' and after this point few recover.

When the long rains eventually broke, large number of the weakened livestock died from pneumonia. The death of so many in the Kachoda valley resulted in a flood of skins onto the market. Prices fell from more than KSh20/– earlier in the year to KSh3/– each. At this time, Adrian visited the far north of the division and stayed with herders of the Ngikwatela section, who whilst were similarly affected by delayed rains, had previously moved some of their herds into southern Sudan. These animals were in much better condition than those which had been kept behind, and few were lost.

In recent years it appears that herders living close to settlements and formal markets are more vulnerable to market forces, as the cash price of goat skins taken from dead animals has been highly erratic and at times collapses. Maize, the cheapest cereal available in the settlements, was sold in 1987 for KSh2.50 per kg and consequently pastoralists were able to purchase only 1–2kg of grain for each skin they sold. Further north, herders were bartering with the Merille and Donyiro to achieve much better terms of trade; up to 20kg of sorghum per skin. Reports from other pastoral areas indicate that herders are frequently manipulated in formal markets, particularly when the pastoral production system is under pressure (Moris 1988). Thus traditional trade may offer better terms, even in times of crisis (Swift, 1988).

A report on drought planning in Turkana written for Oxfam has suggested that local droughts affect some part of the district every three or four years (Swift, 1988). Whilst these droughts may affect local production, it appears that the herders' response is simply to move into other areas. For example, discussions with herders in the Kachoda valley revealed that in earlier drought years many more had herded their animals into the far north of the division. By adopting the traditional response to drought, they had been able to minimize its effects and maintain viable herds. More recently, poorer herders, dependent on food-for-work, have been unable to do this and so are more vulnerable to local droughts. Richard Hogg, in his review, asserted that some development attempts in the district have weakened the traditional drought response (Hogg, 1982).

Apart from this, Moris (1988) suggests that many pastoral production systems are more robust than was imagined until recently and argues that a minimum of three years of declining rainfall is necessary before local drought strategies are undermined. Turkana differentiate between droughts which scorch pasture resulting in the death of small numbers of animals, and *eron* years, 'when everything dies and there is nowhere left to turn' (Swift, 1988). As noted in earlier chapters, however, livestock disease or problems due to raiding are often implicated in these disasters, not drought alone.

A family's ability to recover from a crisis depends very much on the size and nature of the social network to which it belongs. In an extensive network, stock losses might be replaced from outside the affected area. Stock friends and relatives are visited and animals are collected where possible. Following more widespread losses, senior men from a section may also decide to visit neighbouring sections to seek gifts and loans to re-establish small herds. It is interesting to note that in more recent times, the purchasing power of salaried relatives has become a deciding factor in herd recovery. Cash in small stock results in rapid herd building as sheep and goats have faster rates of reproduction in good years. It appears that herds regenerate more quickly after disease than after drought, as animals are left in better condition and large amounts of good quality pasture become available for the remaining animals (Moris, 1988). Finally, large raids may be mounted against neighbouring ethnic groups since many lactating and breeding animals can be gained this way.

Poorer herders take longer to rebuild their herds and flocks and historically some have been unable to re-establish themselves in the pastoral sector. In earlier times such families would have moved out of pastoralism and become absorbed by neighbouring agro-pastoral and agricultural communities. In more recent times, the colonial and independence governments have supported groups of destitutes or ex-herders with famine-relief feeding programmes. Small feeding programmes, which avoid the concentration of large numbers of destitutes, may well assist herders to rebuild their herds as they exchange surplus food for livestock. However, destitute people who remain in a settled camp for long become increasingly separated from pastoral networks. Herders tend only to marry the daughters of herders, and for this and similar reasons, it becomes difficult for long-term destitutes to re-enter pastoral society.

IMPLICATIONS FOR DEVELOPMENT

Although this is only a sketch of livestock herding and sorghum cultivation in the northern Turkana District, a picture of rational and flexible management of resources has emerged. It is clear that Turkana institutions have evolved as an integral part of a sustainable production system. As noted previously, the most questionable aspect of this system, both in terms of sustainability and institutions, is the extent to which the rebuilding of herds after a crisis has depended on raiding.

Some development planners have suggested that it is necessary to build 'herding associations' in Turkana through which resources and development efforts can be channeled. Had such an approach been adopted earlier, it is conceivable that many ghastly mistakes made in the name of development could have been avoided. This said, however, it is symptomatic of the arrogance with which policy makers approach pastoral communities, that they assume that traditional forms of organization can easily be improved upon. Their idea of herding associations might be a good one, but after all, the Turkana already possess herding associations, and at two levels, namely the section (*ekitela*) covering many herders in a large territory and the *adakar*, involving smaller groups.

It would seem that rather than seeking to impose artificial forms of social grouping on pastoral peoples, attempts should be made to understand and strengthen what exists. In Turkana, the institutions most appropriate for a particular intervention will be dependent on the technology involved and may well vary with time and place. For example, livestock vaccination campaigns would probably be best co-ordinated with the leaders of sections (*ekitela*); shallow-well and garden projects could perhaps can best be undertaken by people when they are at their home areas (*ere*); and human and animal-health work might be developed in the context of the smaller herding groups, *adakars*. Whilst composition of these may change from season to season, at least development workers have a valuable reference point around which dialogue can be stimulated and ideas developed.

4. GARDENS AND ANIMAL DRAUGHT, 1985

IMPROVING SORGHUM GARDENS

The food-for-work programme in Kachoda was started in 1981 when the Catholic Diocesan Extension Team encouraged people to move to this hitherto unsettled site. The degraded hillsides of Lokitaung offered little scope for sorghum production and it was felt that cultivation could be better developed at Kachoda to provide an economic base from which people could rebuild their herds.

When the Turkana Rehabilitation Project (TRP) took over the work, so many people had moved to Kachoda that the survey team was overwhelmed, and construction of bunds proceeded without significant technical supervision. Many of the earthworks built were intended to be contour bunds, that is, small embankments built along the contours and spaced between 3 and 20m apart down the slope, depending on the gradient (Finkel, 1985). Over 100km of these bunds were built, but were so badly designed that some never collected any runoff, while others were washed away in the first heavy rains. Because rainfall intensities are high in Turkana, contour bunds are easily breached, and most of those built during 1981–4 have been abandoned. However, some trapezoidal bunds were built in Kachoda during 1984 and seemed to work better (Finkel 1985, pp.12, 59).

With so little achieved in terms of sorghum production, it became clear that once food-for-work was discontinued, all interest in improving gardens would evaporate. Increasing numbers of people were already abandoning the famine camps as herds were rebuilt and it became possible to resume a pastoral lifestyle – and as disillusion with 'development' increased.

On returning to the area in these circumstances in late 1984, Adrian Cullis decided that a strategy based on 'damage limitation' was all that could be hoped for, and started work on two fronts. One was the collection of information based on the experiences of local people, the other was to support the Salvation Army project in the Kachoda area, especially at Manalongoria.

In connection with this second goal, meetings were organized with members of the garden improvement group, at which they were invited to suggest ways of improving the programme of garden development. The response was rather poor. One leader said with great conviction, 'if you want us to remove that mountain, we are able', the understanding being that adequate food would be provided.

It slowly became apparent that the elders main concern was about abuse in food distribution. It seemed that the local site facilitator was taking a percentage of the group's grain allocation. On investigating this, with the support of TRP's area co-ordinator, it was found that twenty bags of maize had gone missing in December 1984. It was decided to transfer the site facilitator to another area, and to make the elders responsible for the distribution of relief food within the group. This agreed, the elders' first commitment was to distribute food twice a week, and thereby reduce problems of irregular distribution which had been affecting poorer householders.

The group then began to share ideas about possible ways to improve gardens. Several people remarked that they had tried to impress upon TRP the importance of levelling gardens, so as to spread floodwater more evenly. It was now decided to start levelling all completed gardens, particularly those which had been planted a number of times, rather than developing yet more sites. Before the work could start, however, there was a need to agree a number of organizational details. Responsibilities were then allocated to group members, elders, and the staff of the Salvation Army project (Table 4). It was also agreed that the group would only receive food-for-work rations whilst improving their gardens, and that after the sites were developed, they would carry out all other work such as maintenance and planting themselves.

Alongside these planning meetings, Adrian continued to meet with elders as a way of maintaining contact and developing ideas. One particularly fruitful area of discussion centred around pastoral movements and organization, where he learned, for example, that the wet season grazing area for herders in Kachoda was in the southern part of the valley,

Table 4   Allocation of responsibilities for work on gardens in the Kachoda area, 1985

Group members	Elders	Project staff
identification of sites	organization of work	design of gardens
site improvements	settling disputes	layout
cultivation	distribution of food-for-work	provision of food
planting	maize	major repairs (caused by
weeding	attending meetings	poor design)
harvesting		
minor repairs		

including the area around Manalongoria. It was therefore understood that gardens could be usefully developed in such areas of wet season grazing, to which nomadic herders would be likely to return, and therefore, to plant. Later, when all the households were interviewed, it was also learned that some members had come from the far north-west of the district. As such, they had limited 'rights' to graze their animals in Kachoda. It was interesting to learn that the elders supported their requests to remain within the group, arguing that all Turkana are 'brothers'. In sharp contrast, however, the elders had denied these households permission to develop gardens, and they were therefore retained in effect as casual labourers. It was proposed that these households be assisted to return to their home sections, and that the elders take responsibility for this repatriation. Despite some disruption caused, and the fact that some only moved to other groups in Kachoda, the move was a good one, as leaders were subsequently better able to represent the expressed views of the whole group.

In order to gain a more detailed insight into Turkana life, Adrian spent some time with pastoral families, and became the owner of a flock of goats. These animals were placed with a family, who benefited from their milk, but he had regular access to them. The head of this family was Lokeris Lorumorr who had worked for the Salvation Army and, together with his mother, had developed a garden at Manalongoria. Lokeris was later employed by the Demonstration Project and has more recently been made responsible for co-ordinating the technical aspects of the rainwater harvesting programme. His life story was recorded in the early part of 1985:

I was born near Manalongoria, the eldest son of Lorumorr's second wife Dida. My mother came to Kachoda from Lodwar as a young girl together with the rest of her family. They were fleeing from the British, against whom her father had fought. My father also was from outside the area. He was a member of the Ngikamatak section from near Lodwar and had moved north with a road construction team.

In my youth I herded my father's cattle in the Lapurr range, but in drier years travelled much further into the Lorionetom mountains. My father had become friendly with a man named Longoria from Uganda, who worked with the colonial service as an interpreter. One year this man planted a garden which yielded well and later became known as Manalongoria. He never planted the garden again because he moved into Lokitaung. But I did not like the town and therefore remained in the Kachoda valley and continued to plant at Manalongoria. Sometimes the sorghum grew well, but at other times several seasons would pass without a harvest.

In 1978 Hillman, the volunteer, saw us planting at Manalongoria. He explained that he had come to improve the gardens and offered our family food to work in our garden. As we were poor at the time, our parents encouraged us to work and my mother began also. I was away herding when Hillman signed on the workers, but I was able to return many times to work. To start with, I was paid a ration of food, but later, because I was so interested in the work, I became the watchman and was given a salary. I knew that if we planted in the old gardens cultivated by my father, we would get a harvest, but it was only much later that I understood that the earth we were moving was to hold more water and make the crop grow better. I was the first person to believe that water could be caught. Most people – even my mother – were only interested in the food.

We tried to plant many crops in the gardens, but have found that only sorghum grows well. We got good crops in 1978, 1979 and 1981. Then 1982 would have been a very good year except the insects ate all the sorghum before we could harvest. The best year was 1981 when I remember we ate sorghum for a long time and gave many sweet sorghum stalks to our friends. I also bought three goats. Since TRP arrived the work has become much more difficult as so much soil has to be moved to build a bund. It is difficult to herd the animals as everyone is working.

I have thought about the gardens a lot since starting work with Hillman and think we can make better gardens. I tried to design my own garden with Paulo Meyen, but the TRP stopped us building the garden properly. I wanted to get more water into my garden and also get more area flooded. I think we could do this on all the gardens in Manalongoria, which at present do not receive enough water. The main problem is that we are not taught how to use the TRP levelling instrument. If I could learn that, I could design a good garden.

WATER HARVESTING DEVELOPMENTS AT MANALONGORIA

With the elders at Manalongoria, it was decided to begin work on three sites, in the hope that at least two would be completed before the time of the rains in March or April. Adrian was able to borrow TRP surveying equipment to survey the sites and mark out bund alignments and areas to be levelled. The elders organized three working groups, and the work got off to a good start, with a high level of co-operation.

Shortly after this, in March, Moshe Finkel arrived in Kachoda to train site surveyors under a NORAD contract and to construct a number of demonstration trapezoidal bunds. Following a preliminary topographical survey, he selected a site at Lochorelupe in the upper part of the Kachoda valley, and drew up a plan for a field layout with fifteen trapezoidal bunds (Figure 9). More than 100 people from camps in Kachoda were contracted to take part in construction and three bunds were finished by 19 April when rains came which satisfactorily 'harvested' water (Finkel, 1985). A number of similar trapezoidal bunds had been constructed in the Kachoda area during the previous year (1984) and sorghum and cowpeas

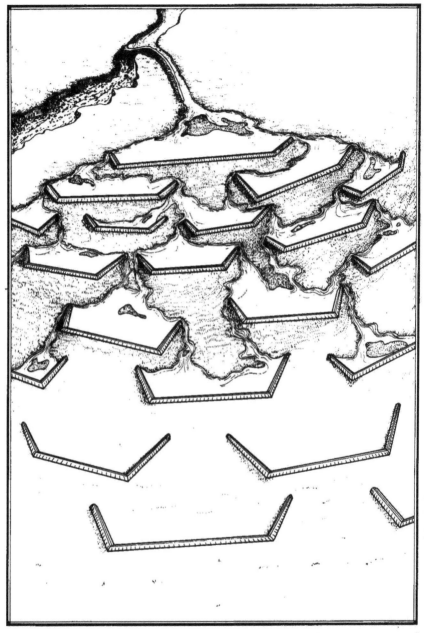

Figure 9   *Layout of trapezoidal impounding bunds for flood diversion and water spreading, showing earth dam, canal and earth bunds. Runoff flooding fills the lower dams when heavy rains fall, allowing crop production over a large area.*
(Source: Finkel, 1986)

had been successfully grown in some of them, while others had been used by livestock for drinking (Finkel, 1985). Thus, for the first time, site facilitators received some training and encouragement and saw some tangible benefits from their efforts.

A small number of staff and elders from Manalongoria also attended the training sessions and took part in construction of four bunds, even though the work was aimed mainly at TRP staff. After the course, Adrian discussed with the group what techniques they had learned which could be applied to Manalongoria. All agreed that the bunds built at the demonstration plot were more robust, better compacted, and therefore less likely to break when flooded than those built at Manalongoria. Several of the staff commented on the good use made of lines for marking out the bunds. Finally, there was considerable interest expressed in the surveying equipment used, which included theodolites and Abney levels. The group recommended that he purchase similar equipment which would enable the project to become independent of TRP. They suggested that he could then design the gardens rather than TRP surveyors, but when asked if they would like to participate in this, they replied that this was impossible as none of them had attended school.

As a result of these discussions, I wanted to introduce simplified surveying techniques which could be used by the group members. The T-frame, which had been used by Francis Hillman, was rather too heavy for carrying long distances. For this reason the line-level was tested and found to be more appropriate to the needs of the group. It was interesting to observe group members, once trained, carrying their line-levels into town, as something of a status symbol. It was possible to build on this and introduce colour coding in order to denote different skill levels. Thus red lines were given only to the most skilled trainees who had completed several training courses. Much of the early training was carried out informally, whilst construction work was being carried out. Later, short courses were organised to which participants from the group were invited to attend. These were held in the afternoons after a full mornings work developing gardens. It was not necessary to offer food incentives to the participants, despite the fact that little or nothing else could be done at the time in the region without food support, as there was a high level of interest. The first course, held in November 1985, was attended by six women.

USING THE LINE LEVEL

Each participant attending a training course was asked to bring two straight sticks about 0.7m long and 10–20mm thick cut to exactly the same length. They were then asked to cut notches near the ends of the sticks, into which a length of line could be tied. The participants then each measured out 10 paces of line and tied the line securely to the sticks. This

Stick with line attached held on bench-mark

Spirit level suspended at centre point of line — Knot

Flat stone serves as bench-mark from which levels are measured

Figure 10   *The line level in use. One person holds each stick so that the line is taut, and a third person watches the bubble in the level.*

was repeated so that the participants could learn not only the method, but also how to train others. The participants were then paired off and with one person holding both pairs of sticks, and the other drawing out the lines, it was possible to find the mid-point of each line. This done, a simple knot was tied at the mid-point to help the user to position the level exactly in the middle (Figure 10).

Having prepared the lines, attention was then turned to the level. Simple 'games' proved useful in familiarizing people with this device. The games were made progressively more difficult until each participant could approach the level, held by another person, and with a single movement centre the bubble. These activities took all afternoon.

The following afternoon, participants were invited to repeat the activities learned the day before and to test each other. They were then divided into teams of three to practice attaching the level to the line at the mid-point. With two people holding the sticks, the third member of the team hooked on the level at the mid-point, checked that the sticks were upright and the line taut. The women were quite self-conscious doing this exercise, as their skin clothing was difficult to manage, and it was therefore decided that they should form separate working groups. When they were more familiar with the equipment, they were happy to work with the men, but some who became project employees eventually purchased skirts for this work.

On the third afternoon the teams were asked to work over a length of rough ground using whatever materials they could find to raise the foot of the stick which the level told them was lowest, to the same height as the other stick. In all cases, the teams chose to level down the slope, building small pyramids of stones and soil as they went. They were less confident going up the slope, which necessitated digging out shallow holes into which they placed the sticks, on small stones. The teams then practised the skill for the rest of the afternoon.

The fourth day included a summary of the first three days' activities with an exercise carried out by each team. The participants were then taken to a site scheduled for improvement, where they assisted staff to mark out the garden, beginning with siting the bench-mark. Large flat stones were used as bench-marks in all gardens to provide the construction team with a constant reference point. This done, the teams then marked out the areas to be levelled. Later, in subsequent training courses, the participants went on to learn how to mark out bunds using pegs and lines, and learned also how to design and build spillways. It is of interest that on one of these courses, the participants requested exercise books in which to sketch out improved designs for gardens.

In this area of erratic and unpredictable rainfall, it was important to ensure the retention of adequate water from a single flood to carry the crop through to maturity. Time was therefore spent with gardeners discussing their experience of improved gardens, which at this time retained different amounts of water on unlevelled sites. Gardeners reported that, provided the water reached half-way up their calf muscle (many held their leg at this point), then sorghum would mature without additional flooding. It was interesting to compare this information with data given by Francis Hillman that for deep soils able to store all the flood water, spillways should retain approximately 30cm of flood water. Data provided by local gardeners was in remarkable agreement. Later, as a result of studying rainfall figures, it was realized that even when gardens were flooded only once in the growing season, there was a strong likelihood of smaller storms which would not produce runoff but which would contribute some rainwater directly to the crop.

It was agreed with staff that gardens should be designed to retain sufficient water from a single flood for sorghum to reach maturity, and that their own estimate of this should be used. A 1.5m stick was then added to the designers' basic levelling equipment which had two burn marks at 30cm and 70cm for measuring the crests of spillways and the tops of bunds respectively. Spillway builders were now able to build up layers of stone to the correct height. Equally important, they were able, using a shortened string, to level along the crest of a spillway to check that the structure would pass water evenly along its length.

By the end of April 1985 it was possible for a visitor to the area to report to Oxfam that 'the Turkana themselves are now surveying, designing and building bunds without outside assistance' (Swift, 1985).

THE LONG RAINS

Conventional irrigation systems can be tested and improved at any time, because water is always available. In Turkana, long delays between the rains result in infrequent opportunities to try out rainwater harvesting designs. During Adrian's time in Lokitaung, the short rains failed each year, and it was therefore only possible to evaluate the technical appropriateness of the structures four times. This being the case, each rainy season was a period of intense anxiety and activity, as Adrian describes.

In 1985, there was a storm with 27mm of rain in Lokitaung on 30th March. I set off for Manalongoria, hoping to watch flood water passing through the gardens. However, on arriving there, I found the sun shining brightly, and no rain. After further abortive visits, I realised that rainstorms at Lokitaung seldom affect Manalongoria. To this day I have never succeeded in arriving at Manalongoria (or most of the other twenty five sites in Kachoda) during an intense downpour.

Not long after this, in April, the gardens at Manalongoria were flooded, without damage to bunds or spillways. Group members began planting, and it was possible to learn how this was done by working alongside the gardeners. During one of these visits, a woman complained that her garden, which was positioned lower in a small series of terraces, was always poorly watered because the flooding was seldom sufficient to provide for all the terraces. Once the problem became evident, attempts were made to incorporate into the design parameters the requirement that each garden should have an inflow from a separate catchment.

Most of the gardeners planted with the traditional digging stick, others with *jembes* (hoes). Between five and ten seeds were planted in each hole, in contrast to the recommended three or four seeds. The latter is intended to achieve an optimum plant population of approximately 8.7 plants per square metre (7.5 per square yard; Powell, 1982). Despite the higher seed rate, however, overcrowding did not appear to result and it was found that most gardeners achieved optimal plant population densities of between seven and ten plants per square metre. It would appear that the higher seed rate compensates for the higher losses which result from using untreated seed. Gardeners also pointed out that if all the seeds were to germinate, they would be able to remove the poorer plants, or transplant them to parts of the gardens with low plant populations.

Visiting other TRP sites along the Kachoda river, Adrian found many gardens that had been damaged by excess flooding. Other gardens lay high and dry, as if no thought had been given to the design of in-flows. By contrast, the demonstration site developed by Moshe Finkel and described above (Figure 9) had worked well. The trapezoidal bunds held useful amounts of runoff water, but subsequent visits revealed that the plots they enclosed were left unplanted. It appears that the ownership of these gardens had not been resolved, and suggestions about communal planting had been rejected.

ANIMAL DRAUGHT

An objective of the work was to introduce animal draught systems, which it was felt would enable gardeners to develop and maintain gardens after food aid was withdrawn. A volunteer had earlier used a pair of donkeys to pull a five-tine cultivator and metal-wheeled cart but neither were used regularly, nor did there seem much interest amongst the local people.

In a completely separate initiative, also at Lokitaung, a local priest had purchased a donkey cart with pneumatic tyres for transporting food to a number of small shops which he was assisting women to establish. As the cart had not been used for some time, Adrian was able to buy it together with the donkeys for hauling stones to construct spillways.

As the cultivator performed well in the light garden soils around Lokitaung Adrian now decided to use it for loosening soil for bund construction, which would otherwise have to be loosened by hand. The cultivator worked well when used for this purpose, but the animals hauling it had crude wooden frame harnesses and rapidly developed large sores on their withers. He was told that the University of Nairobi's Department of Agricultural Engineering had developed a Swiss three-pad collar for donkeys. During a visit to the University, he was able to arrange for two technicians from the Department to travel to Lokitaung and organize a training course for local donkey handlers. He was also able to obtain a set of templates for hames and pad, and an early start was made with the manufacture of harness.

Wooden hames were cut by local craftsmen (Figure 11) using traditional tools and skills, whilst women were involved in preparation of skins and

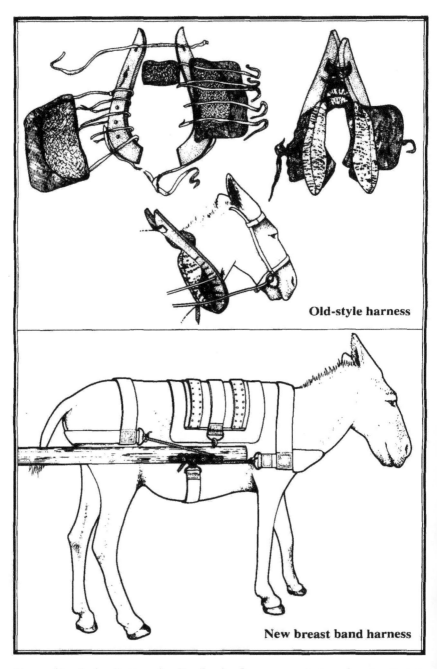

Old-style harness

New breast band harness

Figure 11   *Swiss three-pad collar for donkeys, as made from local materials by staff of the Demonstration Project, and the new breast-band harness.*

hides from which to make pads and bridles. Pads were stuffed with *emoja* fibre (*Sanseveria ehrenbergii*), prepared like sisal by beating the green stems and later washing to leave only cleaned fibres. Collars and harnesses were thus manufactured for between KSH140 and 150/– or approximately $10. Other equipment later made by local craftsmen included sledges for carrying soil and stones, Indian standard design scraper *baords*, and Ethiopian *ards* for soil loosening and small dam scoops (made from fuel drums).

Some useful discussions were held with members of the group at Manalongoria about their perceptions of draught and pack animals. Donkeys are the local beast of burden and there was no antipathy to them being worked in the gardens. Traditionally, donkeys are used for carrying household goods during migration from one herding camp to another. They also carry small children and young kids and lambs during migrations. Where long distances have to be covered, donkeys are employed for water-carrying with five gourds on each side of the pannier, and their use in the sorghum trade with neighbouring ethnic groups has already been mentioned.

Once the project donkeys were trained and earth-moving equipment manufactured, it was possible to start donkey teams working on construction sites. At first, staff worked the donkeys but later encouraged members of the groups to handle them, and began informal training courses in driving, turning, stopping and use of the equipment. After the donkeys had been working with a group for three or four weeks, a discussion was held to share experiences. The group members were quick to point out the improvements associated with the new harness, and the increased work rates. They concluded that the donkeys should continue to work in their gardens. Despite enthusiasm on this score, however, suggestions that the group members train their own animals for this purpose were met with steadfast refusal. Some pointed out that their donkeys had been lost in *Lopiar*, whilst men known to own donkeys responded that women could move earth faster than donkeys, and they saw no reason to use their animals.

Adrian had hoped that this resistance would dissipate when it was demonstrated that the animals suffered no ill effects from regular work. However, such experiences seemed to have no impact. Herders were totally opposed to working their own donkeys.

I tried to promote draught-donkey systems through restocking and incentive payments. In one scheme, it was proposed that groups purchase donkeys with maize set aside from the garden contracts. These animals would be allocated to a group member without donkeys, and in return, the owner would agree to work the animals for a set period of time. Whilst half-a-dozen animals were purchased in this

scheme on a trial basis, resistance was met from people who
were unwilling to allocate their maize rations for the purchase of
donkeys.
   After other suggestions were rejected, and it had become clear
that the project staff were also unsympathetic to what was being
attempted, it seemed that the only way to introduce animal draught
was to purchase donkeys and establish a demonstration herd.
However, recognizing that if I attempted to purchase donkeys myself,
prices would be increased exorbitantly, it was agreed that a staff team
of buyers should go south along the lakeside, to buy animals.

   On the first trip, the staff team were given funds to purchase twelve
donkeys, and in addition, three male camels for the establishment of a
pack-camel team. They returned six weeks later with eleven donkeys
(including one female), two camels and the tail of a third which had 'died
on the journey home'. Although it was strongly suspected that the camel
had been slaughtered for a meat feast, it was impossible to confirm this
until much later. Further trips organized by different members of staff
were more successful and the project quickly built up the donkey herd to
the target number of thirty. In order to provide the donkeys with
supplementary forage in the dry season, efforts were made to procure
some 2000 bundles of sorghum stover from gardeners in Kachoda. About
half this quantity was satisfactorily purchased, but it was only possible
to obtain the full amount by buying from gardeners along the lake
shore. Transport proved a difficulty there, and it was necessary to
hire TRP lorries. Thus the resources needed to sustain the donkey draught
project rapidly exceeded anything that local groups could afford or
manage.
   Once the demonstration herd was established and trained, gardeners
could borrow donkeys from the kraal where they were kept, work them for
up to two and a half hours (during which time they were to be rested for
half an hour), and then return them to the kraal. People were keen for the
donkeys to assist in the levelling work which was found to be particularly
demanding. People with gardens on the flood-plain also used them to
collect stones from the neighbourhood hills. With an increasing number of
donkeys ready for work, the project craftsmen were fully employed
producing earth-moving equipment and effecting repairs. Simple sledges
overlaid with sacks to hold soil were used extensively to carry soil to
areas that needed building up. When almost enough soil had been dumped
by this means, scraper boards were dragged across the gardens to fill
in shallow depressions. Later, earth scoops replaced the sledges (Figure
12).
   The manufacture of earth-moving equipment in Lokitaung proved to be
difficult as there were few local sources of metal and other basic materials.

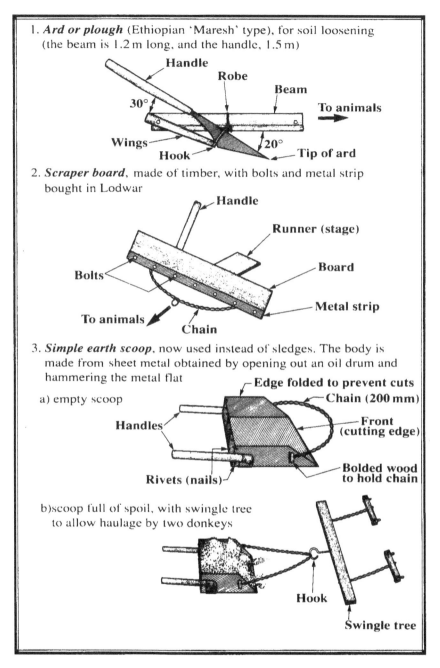

1. **Ard or plough** (Ethiopian 'Maresh' type), for soil loosening (the beam is 1.2 m long, and the handle, 1.5 m)

Handle
Robe
Beam
30°
To animals
Wings
Hook
20°
Tip of ard

2. **Scraper board**, made of timber, with bolts and metal strip bought in Lodwar

Handle
Runner (stage)
Bolts
Board
To animals
Chain
Metal strip

3. **Simple earth scoop**, now used instead of sledges. The body is made from sheet metal obtained by opening out an oil drum and hammering the metal flat

a) empty scoop

Edge folded to prevent cuts
Chain (200 mm)
Handles
Front (cutting edge)
Rivets (nails)
Bolded wood to hold chain

b)scoop full of spoil, with swingle tree to allow haulage by two donkeys

Hook
Swingle tree

Figure 12   *Earth-moving equipment for use with donkey draught, designed for manufacture by the staff of the Demonstration Project. (Source: based on drawings by Simon Barasa)*

It was found necessary to import tools, finished timbers and much of the metal from Kitale, 500km to the south. A return journey to Kitale took three days and therefore it was not possible to make regular visits. Transport problems were not the only obstacle, however. Lack of experience was a constraint when it came to designing light-weight equipment which was strong enough to withstand the heavy work in which it was used. In early 1986, an agricultural engineer was contracted to carry out a short consultancy.

REAPPRAISAL AND REORGANIZATION

It had always been Oxfam's intention that the Demonstration Project should include an example of Yemeni-style spate irrigation. Requiring earthworks on a much larger scale than the sorghum gardens, it was envisaged that the animal draught needed to construct them would involve oxen rather than donkeys. Aware of the difficulties with introducing donkeys, Adrian was doubtful about these ideas. Furthermore, the scale of a spate irrigation system seemed too large to be managed through Turkana institutions. Even so, he helped the Water Harvesting Co-ordinator for the District search for sites of around 200ha on flat land near suitable water courses from which flood flows could be diverted. They rejected the Kachoda valley as too dissected, but found a large open area in the Kotome valley which seemed suitable for improved grassland production.

Later they were told by herders from the Kotome valley that there was little point in improving grassland there because they were always forced to move their stock from that area early in the dry season due to shortage of water long before there was any shortage of grazing. Despite this, work on the spate irrigation project was being pushed forward on the basis that it would at least demonstrate the principle. There was a suggestion that TRP should close its food-for-work centres in Kachoda and encourage the labour force to move to the Kotome valley site, and it was also suggested that Adrian should move in order to supervise the work.

I found myself in increasing disagreement with these developments, and not only refused to move to the Kotome valley, but persuaded TRP not to close their Kachoda centre. But Oxfam's Nairobi office was now receiving complaints from TRP staff that I was stepping outside the terms of reference agreed for the Demonstration Project. To try and resolve these differences, a meeting was held between Oxfam and TRP in September 1985 at which I argued strongly that mounting a demonstration was not sufficient to guarantee the participation and interest of local people. The Lorengippe project had demonstrated spate irrigation in the 1960s, and TRP had been demonstrating other water harvesting methods new to Turkana since 1981, but local people had not adopted these techniques for themselves.

It was necessary, I stressed, to move more slowly and learn from local people, meanwhile strengthening the ability of existing pastoral institutions to manage their own lands.

There was thus a clash between two different concepts of how people are stimulated to adopt new technology, one stressing demonstration and imitation, and the other based on a process of dialogue. In an effort to reconcile these divergent views, based on different philosophical view-points, the meeting agreed that the work on Kachoda could continue, but that there should also be a larger-scale spate irrigation demonstration.

Shortly afterwards, Oxfam's arid lands consultant, Brian Hartley, came to Lokitaung to review differences still outstanding. Although he was committed to the idea of spate irrigation, he wanted to see this technique used for sorghum production, not for improvement of grassland, and he was keen to develop a potential spate irrigation site of 25ha at Loarengak which he had identified in 1982.

When Brian Hartley volunteered to establish this project, I agreed since I could then disassociate myself from spate irrigation and ox draught whilst allowing the original terms of reference to be observed. Faced with increasing pressures, I was anxious to find ways of satisfying Oxfam's requirements which, at the same time, would allow the work started at Manalongoria to continue. But I was also aware that neither Oxfam nor TRP were really convinced by my arguments, and that the work in the Kachoda valley would have to expand and show results if they were to be persuaded.

About this time, the Salvation Army ended its connection with rainwater harvesting in the Lokitaung and Kachoda areas, partly because Oxfam did not renew a grant, but also because its own Nairobi headquarters felt concern that the project was still unable to support itself after seven years. It thus became possible for the Demonstration Project to recruit the Salva-tion Army's water harvesting staf, except that Paulo Meyan, the most experienced member, was leaving the area on being appointed Assistant Chief of Loruth sub-location in his home Ngikwatela section. With his departure, there was no obvious leader, and it was decided to give the four other staff members separate responsibilities and encourage them to work together as a management team. One was to liaise with garden owners; another was to administer the food-for-work programme; a third was to be re-sponsible for the donkeys and animal draught; and a fourth, Lokeris Lorumorr, was to design improved gardens and supervise construction. The overall aim of the work agreed with this team was to complete improvements of ten or twelve gardens on a range of different sites before the long rains of 1986, thus enabling the project to convince sceptics of the value of a partitipative approach.

One of the team, Ekitela Eleman, was a local elder. As custom dictated, it was agreed that he should take on overall responsibility. Finally, it was decided that the management team should meet weekly to review progress and to share problems. The meeting ended after Ekitela had organized the following week's activities and suggested that he would visit two sites further south along the Kachoda river. The apparent ease with which the group assumed responsibility for developing the work in Kachoda served as a useful reminder of the depth of resources which could be tapped from local people.

As work progressed and additional groups were identified, the weekly meetings became an important forum for discussion. Sometimes other members of staff and leaders of family groups were invited to participate in these meetings, in order to draw on relevant experience from outside the group. One of the early criticisms voiced at the meetings was the delays faced by people waiting for Adrian to visit work sites and calculate the amount of food to be allocated for work done. This led to the idea of a contract system. Gardens were marked out and the amount of earth moving was estimated in advance of work starting. It was then possible to agree with the group the total allocation to be made for developing the garden, and that amount was made available at the store. The group was now able to draw down on this food as and when they wanted, provided that a small number of bags were left until the work was completed and checked by Lokeris. The new system was introduced for a trial period and was later fully adopted. From November 1985, more than 100 contracts had been made in three locations, and in only three or four was the work abandoned or the contract withdrawn.

EXTENDING TO NEW SITES

Extension of the work in this situation entailed Ekitela in seeking collaborative arrangements with existing TRP sites in the Kachoda valley. One was at Lomareng, 5km south of Manalongoria, where the first contacts had been made in July (Figure 14 below). In October work began at Nadiye nearby. Two other sites were added in November and December. Then a further development, in December 1985, came from people interested in improving gardens at Milima Tatu, 60km from Lokitaung on the road to Lodwar. Work on this site eventually encouraged the establishment of a second administrative centre for the project at Kaalin, but ironically, the Lotorup gardens were later abandoned.

Site development took a considerable amount of time, as a standard 'blueprint' was not considered the best approach. Whilst it was always possible to construct ten, fifteen, or even twenty trapezoidal bunds on a carefully selected demonstration site, few groups of gardeners chose land where such a layout was possible. Often, there were a whole range of

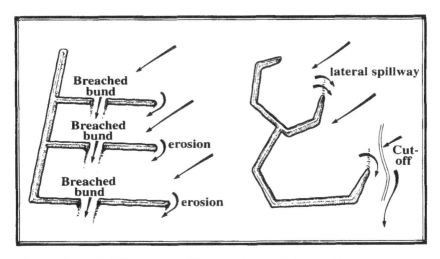

Figure 13   *Rehabilitation of a TRP garden at Nadiye, with arrows indicating the direction of runoff water flows.*

design difficulties to think through, including existing TRP bunds, dissected catchments, inadequate inflow or slopes steeper than assumed by the blueprint. It was impossible to abandon all these more difficult sites, and it seemed clear that each one should be approached individually, working with the natural topographical features rather than against them. Training the team to adapt design principles to meet the requirements of each site was itself difficult, but gave them a better approach than if they had merely been shown how to reproduce a standard design.

At the end of the year, the project was developing several sites with bunds laid out on a variety of plans. Some had contour bunds (terraces), and others modified trapezoidal bunds. Some gardens were on new sites, but many had previously been developed by TRP and were in need of redesign and rehabilitation (Figure 13).

Once design work on a particular garden had been completed, the amount of earth to be moved had to be quantified. This information was then transferred to a contract form, and the food allocation calculated. The garden owner was informed of the amount of maize which could be used during the construction programme and presented with a store card to take to Pius Chuchu, the project monitor and store clerk. He retained the card at the TRP store, and as the sacks of maize were withdrawn, he would register the number and calculate the balance. Details of seven contracts are given in Table 5. Much of the work, particularly land levelling, was done with the assistance of donkey-hauled equipment, as described previously.

Table 5 Details of contracts between the Demonstration Project and seven groups of gardeners drawn up during 1985. The size of the groups varied, and the time for completion of each contract was adjusted to allow for this.

WORK GROUP	WORK TO BE DONE			PAYMENT	DURATION
	Levelling (m³)	Bund construction (m³)	Length of spillway (m)	(bags) maize)	Time for completion (months)
1)	111	450	10	20	5
2)	386	230	only repairs needed	21.5	3
3)	309.5	469	10	28	7
4)	400	405	10	29	2
5)	606	1130	lateral spillway	61	5
6)	574	540	8	38	6
7)	800	750	–	53	6

Meanwhile, the ox-draught experiment at Loarengak was under way, and Adrian had to make regular visits because Brian Hartley had completed his contract to set up the project. Work on the spate irrigation site was complete, and the oxen were now employed improving traditional sorghum gardens in the vicinity.

With new work in hand, I had neglected the investigations about local institutions and the 'base-line' survey I had begun with local people. Some of what I had intended was now being done in a different way through a survey of the women of Manalongoria, who formed 70 per cent of the work-force on garden sites (Watson 1988).

With the appointment of a new Oxfam field director in Nairobi later in 1985, survey work was however given more encouragement than previously. It was therefore suggested that local 'monitors' should be recruited to collect detailed information on a number of topics, including pastoralism, sorghum production, and the impact of food-for-work. Some personnel were seconded from TRP to carry out these tasks.

MID-TERM PROSPECTS

The beginning of January 1986 was the half-way point in the two-year life planned for the Demonstration Project. One very disturbing event occurred soon after this date, after Adrian had been absent for a month due to illness.

On returning to work, I was greeted with the news that one donkey had died and several more were on the point of death. I travelled to the donkey kraal and was distressed to find that another donkey had died and four more were so seriously emaciated that they seemed unlikely to recover. In fact, five died in total. I called a meeting of project staff in Lokitaung, to underline the importance of the meeting.

I was quickly confronted with a number of painful truths. Firstly, I was told that local groups considered the donkeys to be the property of the 'foreigner', and were therefore little concerned about their welfare. The staff explained that although they had spoken with group leaders, they had been unable to prevent people overworking the donkeys. They said that some of the animals had been worked for five hours at a stretch, and in two cases, donkeys had collapsed while in harness.

Having learned the basic facts, I then challenged the staff about their responsibilities. A number of them pointed out that they were not directly involved with training donkeys or looking after them. Those who were responsible claimed that they had been detailed to train donkeys, and train gardeners in their use, but not to regulate the work programme. It was impossible to get any staff member to admit responsibility.

On the credit side, however, it was pleasing to find that the construction programme had been going ahead smoothly in his absence. Several gardens had been completed, and the design team took the initiative in opening new sites despite the fact that they were unable to estimate the amount of work involved in terms of cubic metres of soil. The groups had therefore been allowed to withdraw up to three bags of maize, which they were aware would be deducted from the final contract.

In February 1986, a consultant from Britain, Jeremy Swift, was able to confirm, as part of a mid-term review, that Oxfam's terms of reference were being fully met. However, he pointed out that although improved sorghum gardens could produce a good yield (1000kg/ha) in perhaps two years out of every five, this could not provide the whole livelihood for any group of local people, not least because of the seasons when no cropping was possible. Improved sorghum gardens should not, therefore, be seen as a means of support for people made destitute by earlier famines, but as a means of complementing pastoral livelihoods. The aim should be, 'to diversify the pastoral economy, not create an alternative agricultural economy' (Swift *et al.*, 1986, p.4).

The target population should not, therefore, be destitute people, but poorer pastoralists, whose livestock did not provide a secure livelihood. The review went on to suggest that the project could best achieve its objectives by assisting TRP to develop better managed small-scale sorghum gardens. It was argued that the objectives of the project should be redefined to focus on the support of pastoral households, and that the project experience be extended throughout Turkana by teams of specialists. Finally, the review suggested that a number of complementary activities be added to the work, including institution-building, restocking and browse regeneration through co-operation with other projects (Swift *et al.*, 1986). It was thus proposed by the review document that in the second year of the project, 1986, project staff should train three teams of specialists to develop work in other locations. While one team would remain in Kachoda, two others would be located further from Lokitaung under TRP technical co-ordinators. It was also suggested that monitoring be made a priority; it should include plot record cards and 'qualitative data on the life of water harvesting groups'.

The review document was presented to Oxfam and ITDG in April 1986, and stimulated a considerable amount of discussion. A number of modifications were agreed which redefined objectives for the second year in a way which was more in line with Adrian's own ideas. The project would begin to broaden its activities whilst retaining effective management control and therefore, it was suggested that the project should become independent of TRP. He was to retain effective managerial control over the expanded programme, whilst devolving more responsibility to local groups. The project would continue to work through home area (*ere*) groups and garden groups, and would seek to support and strengthen these traditional institutions rather than establishing new and artificial associations. It was expressly stated that additional programmes would be developed at the pace set by these institutions, and the project manager would not seek to impose developments upon them.

5. NEW LOCATIONS AND EVOLVING MANAGEMENT, 1986-7

GEOGRAPHICAL EXTENSION

By the middle of 1986, the Turkana Water Harvesting and Draught Animal Demonstration Project was helping people develop improved sorghum cultivation in three locations, namely Kachoda (where the work started), Kaalin (to the west) and Loarengak (on the lake shore). There had also been a short-lived garden scheme at Naramum, to the north of Kaalin in the Kotome Valley (Figure 14).

People in the three main locations came from two different pastoral 'sections', the Ngisiger and the Ngiyapakuno, and as work progressed, it became clearer that attitudes differed considerably between them, not least because their traditional lifestyles had been different. The Ngisiger, particularly those on the lake shore, had a much more mixed economy in which sorghum cultivation had long been a part, while the Ngiyapakuno were more purely pastoral. Indeed, most Ngiyapakuno people (like the Ngikwatela further north) were contemptuous of cultivation, preferring to obtain their cereals by trading.

Kachoda was a settlement established by TRP initially as a food-for-work camp and was on the border between these sections. The people with whom the Demonstration Project had begun work in 1985, as described in the previous chapter, were mainly from the Ngisiger section. Although they had less of a tradition of sorghum gardening than the Ngisiger on the lake shore, in the twenties or thirties some of them (or their parents) had planted gardens in an area at the northern end of the Kachoda valley known as Lomogol. This was later abandoned because of the security situation, and the southern part of the valley, which the people now used more intensively, had fewer sites suitable for gardening by traditional methods. For this reason and because of dislocations caused by the establishment of the Kachoda famine camp, the area was not proving ideal for a rainwater harvesting programme.

In mid-1986, the Demonstration Project entered a new phase, following the review described in the previous chapter. On the one hand, efforts were made to develop rainwater harvesting in new areas, away from Kachoda. On the other hand, there was an attempt to monitor the work and record results more effectively.

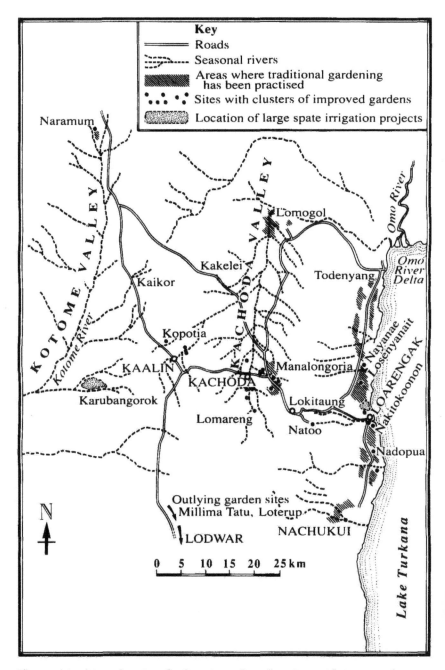

Figure 14 *Map showing the location of garden sites with improved rainwater harvesting systems built between 1985 and 1990.*

As mentioned previously a large spate irrigation project had been begun in the Kotome valley west of Kachoda and Kaalin, by TRP. A site had been chosen on a wide, open plain, and it was proposed to divert the whole of the Karubangorok River onto the plain and to spread the water by a series of banks. The proposed earthworks were massive, and for this reason, large numbers of people were needed to work on the scheme. However, conditions were appalling, partly because of drinking-water shortages. In late 1985, Paulo Meyen, who had worked with the Salvation Army in Lokitaung, came to live in the area after being appointed assistant chief for Kaikworr location. He soon became convinced that the scheme was poorly conceived, and that, as a forage improvement project it offered little benefit for local people. He therefore encouraged people to think about alternatives to the work, and approached the Demonstration Project for support to develop sorghum gardens in another part of the valley.

The place selected by the elders was at Naramum, 45km to the north, where they remembered several families growing sorghum successfully thirty years or more earlier. The area planted was a delta fan, over which a seasonal river cascaded off the mountains. On this site, water-spreading occurred naturally. Some five groups of nearly ninety adults moved there in early 1986, and established a camp with meeting centre and dispensary. Because sorghum had been grown here in the past by traditional Turkana methods, and because there was considerably more rain than further south, grass strips along the contours rather than earth bunding were recommended, and more than twenty gardens were laid out. Unfortunately, though, at the time of planting, Toposa people from southern Sudan attacked herders just north of Naramum, and all the herds were moved far to the south. The garden area was then exposed to raiding. People evacuated it, and the site was abandoned. Even so, this had been an instructive experience, because it had shown that people were prepared to organize their own programmes and tolerate a degree of hardship in experimenting with their own development.

Except for this brief episode at Naramum, the first new location for water harvesting work was Kaalin, in the territory of the Ngiyapakuno section. An opportunity arose here when one of TRP's mobile extension team, a man named Akorot, retired from his TRP post and settled in Kaalin. He was ideally suited to the work, and it was largely to his credit that a programme was established at Kaalin. However, people here had even less of a gardening tradition than those at Kachoda, and if the implications has been appreciated earlier, the area of rainwater harvesing might not have been chosen.

This said, however, studies carried out by a member of staff indicated that many families in the area had semi-permanent home areas (*ere*) along the Kopotia and Kaalin rivers, which offered permanent water from wells sunk in their beds. Here, in the dry seasons, the old, young, pregnant and sick were living off small milk herds, whilst the breeding herd went off northwards to the dry season grazing grounds. Thus there was an established social system which would enable local people to take greater control than usual of any work which was initiated.

The last of the new locations at which the Demonstration Project began to work was the lake shore around Loarengak. There was a strong tradition of sorghum gardening here, and in general conditions were more favourable for rainwater harvesting than at either of the other locations. Although some work had been done in 1985 in connection with the spate irrigation project begun by Brian Hartley, it had seemed best to leave the area to TRP. However, in April 1986, opportunities for working with traditional gardeners were looked at again. It was an ideal area, not only because of the long-standing agricultural tradition and stable *ere* groups, but also because there are areas of flat, fertile land which in times of rain become inundated as runoff water is prevented from reaching the lake by a barrier of sand-dunes.

Evidence of the local sorghum cultivation tradition was especially clearly expressed at a meeting where Pius Chuchu, now project secretary, spoke about it as follows:

My grandfather was the first person to plant at Nakitokoonon. He was called Lopurr. He planted there together with his family, relations and herding friends. One year they did so well that the Merille came to trade for sorghum. They exchanged sorghum for livestock and became rich. They then left off planting as they took their herds northwards to better grazing. About 1962, Lopurr's first wife and son (Pius's father) returned to plant with some close relatives and the wives of some of Lopurr's age-set, with whom he had been herding. They planted for three or four years. After this they returned north again, and established themselves at Todenyang. My uncle, however, remained in Nakitokoonon, and they remain there still.

In those days the Turkana could plant any place they wanted to. All that had to be done was to tell the elders and they would inform the herdsboys, who would prevent animals from eating the crop. Good sites could be identified easily, as sorghum which had split or been dropped the last year would spring up. This sorghum was by right the property of the camp which had left.

After leaving Nakitokoonon, my grandfather and father also planted at Nayanae Losenyanait. If I wanted to plant, I would return there and ask the elders who remember them ... it is closer to the family herds than is Nakitokoonon.

MONITORING, STAFFING LEVELS AND GENDER ROLES

At the same time as the Demonstration Project was extending its activities to these new sites, an improved system of data collection and record-keeping was being set up with the help of Ambrose Ndirangu, a former priest with the Diocese of Lodwar, and Adrienne Martin, a social scientist from Britain, who came to Turkana as a consultant for three weeks in September 1986. She suggested a rationalized system of information collection, and helped train the staff who had been recruited to work as monitors under Ambrose Ndirangu's leadership (Martin, 1986).

Changes in data collection and project management made as a result of this work included a new procedure by which people applied for support to improve gardens. Recommendations were also made about the collection of data on the size of fields, crop yields, and runoff events. An effort was made to identify relevant Turkana categories, in order to ensure that the monitoring team wherever possible recorded data in terms of known Turkana classifications, notably with respect to plant names (Morgan, 1980) and soil types.

The monitors were given some initial training in the use of these various methods of data collection. They found the collection of social science data particularly difficult in interviews or structured discussions as they struggled to understand the value of this information. For this reason, they tended to hurry the conversation along, missing opportunities to pursue revealing comments. However, none of the team had previously collected social or agricultural information. Also most had received only primary education. Even so, they worked hard and employed their considerable local knowledge to good effect, collecting a good deal of useful factual material. Their work laid the basis for a more questioning approach which helped the staff develop further a flexible and responsive project.

Little was known in the early stages of the project about the managerial and organizational capacity of local staff. The project had therefore employed a considerable number of people, each with a very specific role. As staff demonstrated their technical and organizational competence, it became less necessary to employ such large numbers, and fewer staff were taken on at the new locations. This left Kachoda with more than three times as many people employed to do the work as at Kaalin and Loarengak, even though duties were similar. This caused some confusion and resentment, which made it necessary to introduce a more structured staffing policy. After consultation, it was decided that the following five paid employees would be employed at each location:

○ an *ekarabon* or elder, who would be both an extensionist and the team leader, responsible for developing and co-ordinating the work at local level and for working with the elders in the home areas (*ere*).
○ a monitor and store clerk, appointed to work with the team leader as his 'pen'.
○ two rainwater harvesting designers, to develop improved gardens and to train *ere* gardeners in construction techniques.

o an animal draught *fundi* (technician) to train people in the working of draught animals and to make equipment.

Although implementation of this staffing policy without delay was intended, it took some time to achieve the specified level of staff because the people currently employed were all performing well, and it was understandably difficult to decide who to retain. However, in mid-1987, there were four extensionists and four monitors distributed between the three locations as well as designers and *fundis* in approximately the numbers indicated (Martin and Gibbon, 1987), and by the end of the year or soon after, the total staff had been reduced to sixteen, of whom five were women (Watson, 1988).

The latter figure highlights another problem. It was obvious from the outset that there were considerably more women than men in the food-for-work camps, and in the groups developing gardens through rainwater harvesting. A registration at Manalongoria in May 1985 had indicated that more than 60 per cent of members were women, and a survey of ten water harvesting groups at Kachoda, two at Kaalin, and one on the lake shore in late 1986 showed that, in each location, only one-third of members were men (Martin and Gibbon, 1987). Yet it was initially found much easier to engage men in discussion and decision-making, and more difficulty to involve women, and recruitment to the staff reflected this.

As more information became available about the traditionally important role of women in cultivation, the Demonstration Project was increasingly able to improve opportunities for women in skills training. This was done slowly, with firstly 25 per cent women trainees on informal courses, and later 50 per cent. By taking a slow approach, it was possible to make progress without alienating the men. There was no problem with acceptance, and 40 per cent of trained group members are now women. In particular, many of the members who have been trained to work with draught animals and function as animal draught technicians at *ere* level are women (Table 6). A deliberate effort was made to achieve this, because it had been pointed out that otherwise the introduction of animal-draught technology would have been a means whereby men would have come to dominate all project work (Watson, 1988). In order to maintain gender balance in the training of '*ere* technicians', it was necessary to have some special training courses for women, but this was helped by the fact that when Brian Hartley first brought draught oxen to Loarengak, a woman named Lokarach Ikwachu volunteered to work the animals. She had learned how to plough with oxen in Karamjoa (Uganda) and came to play a vital part in project work in this area. She is still employed as an animal-draught technician in Kachoda.

The most important point of contact between garden groups (*ere* groups) and the project was through the extension work of the *ekarabon* or elder.

Table 6   Number of people trained to work with draught animals up to July 1987. (Source: Martin and Gibbon, 1987, p.56)

	Women	Men
FUNDIS		
(animal draught technicians)	–*	3
GROUP MEMBERS		
(people in water harvesting and garden improvement groups)		
Loarengak and the lake shore		
trained with oxen	10	8
trained with donkeys	3	–
Kachoda and Lokitaung		
trained with oxen	–	3
trained with donkeys	12	10

* The senior animal draught technician, Lokarach Ikwachu, is not included since she was not trained by the project.

An experiment was therefore made in appointing women to this position with a view to encouraging women gardeners to seek help in developing gardens in their own right. Two very able women were selected and trained, but both found difficulty in visiting some of the more remote sites, and some people were unwilling to allow the women to become involved in either conflict resolution or signing contracts. The mistake was perhaps to expect the women to take on the male elder's role rather than exploring ways in which the role could usefully be redefined in relation to the project goals concerning women gardeners.

Once it was agreed that the programme would be developed within a context of *ere* groups, it was also decided that the number of such groups in each location should initially be limited to twelve. This was to ensure that the staff could properly supervise the work and carry out adequate training. The staff teams in each of the three locations were encouraged to meet together at least once a month to plan their work, and after a while these meetings were enlarged to become informal committees. As part of a broad process of local control and decentralization, one man and one woman from each *ere* or home area were invited as *ere* representatives. These new committees then replaced the meetings that Adrian Cullis used to convene.

CONTRACTS AND METHODS OF WORK

The expansion of the Demonstration Project to include work in three locations meant that local staff had to take on more responsibility for the day-to-day programme of work. For the technical staff, this represented a continuation of a process started much earlier. Several had first been appointed in Kachoda, and had gained experience there before being transferred to Kaalin or Loarengak.

For the team elders (extensionists) who had now become leaders, there was however a rather more rapid increase in responsibility. It was therefore thought that they would benefit from a simple policy framework to guide their work. This would include criteria to help them decide which households to support in improving gardens. For example, all such households should have well-established rights in the area, preferably (but not necessarily) including rights to cultivate, as well as grazing rights, ancestral graves, and so on. Secondly, any household assisted should own at least fifteen female goats, and hence be committed to remaining in the pastoral sector, and thirdly, it should be able to organize a team of workers to develop the garden numbering up to twenty-five people. It was envisaged that staff would no longer be involved in organizing such work groups.

Other considerations included a reduction in food-for-work rates. Contracts would not, in future, exceed a total of twenty-five bags of maize, irrespective of the size of the garden, or the amount of earth to be moved. Moreover, it was now suggested that households should make a small contribution to the project as an *eboka* or thanks payment, usually in the form of a goat. In addition, people would have to appreciate that contracts were binding, but that they were also of limited scope, covering only the construction of earthworks. Once a garden had been completed, there would be no additional bags of maize to pay for its repair or maintenance.

Previously, contracts had been made with *ere* groups, and the decision was then made within the groups as to who should establish planting rights in the gardens. Under the new policy, it was necessary to know in advance who would become the 'owner'. Contracts were now made with individual households, making her/him responsible for their own gardens. However, the project did not want to see existing *ere* groups break down, and therefore contracts were only made after consultation with elders in the area. Interestingly, as these changes were introduced, some *ere* group members transferred to neighbouring TRP groups. It was evidently too hard for them to face making their own decisions, accompanied by reductions in food-for-work, after so long in the camps.

In September 1986, the project was still dependent on food supplied by the World Food Programme, and payments to people working on their gardens had to conform with TRP food-for-work rates. These were 2.5kg of maize plus some oil and beans for every cubic metre of earth moved. When relatively few gardens were being improved, it was quite simple for Adrian to visit each one to calculate the amount of earth to be shifted and therefore, to establish the amount of food required. There were few delays and most gardeners seemed to prefer to have a foreigner calculate the food-for-work rates. It was said that foreigners were more honest and there was no bribery. However, as the project developed, it became more and more difficult for him to stay ahead of the work, particularly when some visits resulted in suggested changes in design. There was soon an increasing number of complaints about the delays, and it seemed best at that point for local teams to administer the contract system.

Local staff could prepare a contract by measuring the outlines of the proposed earthworks, completing a work-sheet, and using a ready-reckoner to calculate the volumes of earth and the quantity of maize to be paid out. A store card was then completed which recorded the amount and dates when maize was withdrawn. Staff followed this up with regular visits to each site to assess performance and to ensure that the food resources were being used properly and not finished disproportionately ahead of the work. After this system was introduced, a total of 175 gardens was completed with only five abandoned during construction. In order to make it as certain as possible that construction was completed, including the finishing touches, it became the practice to retain part of the contracted food grain until all the work was done.

During the latter part of 1986, TRP's allocation of food aid was substantially reduced, and food-for-work opportunities were restricted. Food continued to be available to meet the Demonstration Project's contracts however, and with less work available from TRP, improving gardens under contract became an increasingly attractive option. There was, however, an increase in the demand for gardens, particularly in Kachoda, close to the larger food-for-work camps. However, it was not the aim of the project to function as an employment agency, and they therefore decided to reduce food-for-work payments from November 1986 (see Table 7). This resulted in a good deal of criticism and complaint which had to be handled by staff in their local committees. However, the staff soon developed the capacity to handle criticism, and were able to point out that, as many families had been able to rebuild their herds, the time for food-for-work was drawing to a close.

Whilst all went reasonably well on this front, it was clear that reduction in the availability and amount of food was hurting the poorer households and particularly, some women-headed households. Some women had depended on hiring out their labour to other people who were building earthworks to improve their gardens. Many of them could no longer find work, as family labour was increasingly used in construction in order to keep the reduced maize payments within the family.

The reduction also resulted in a substantial slowing of the work at Kachoda. For some time, there was a steady stream of people away from the project as families moved off with their herds. Some people later abandoned their gardens and moved out of the area altogether. This trend, as we will see in the final chapter, has subsequently slowed, and some gardeners have more recently returned to re-establish themselves as

Table 7 Plot construction and food allocation trends, showing the reduction in food-for-work payments per plot from November 1986. After January 1987, changes in the design of gardens led to less levelling (because sites on gentler slopes were chosen) and fewer long lengths of spillway (because of the more widespread adoption of lateral spillways). (Source: Martin and Gibbon, 1987, pp.42–3)

PERIOD	Number of plots improved	Length of spillway per plot (m)	EARTH MOVED PER PLOT Levelling (m³)	Bunds (m³)	FOOD ALLOWANCE PER PLOT Maize (90kg bags)	Oil (4 gall. cans)
KACHODA						
Oct 85–Apr 86	10	21	454	469	32.5	20
Jun–Nov 86	10	10	438	381	26.1	22
Nov–Dec 86	7	5	458	393	11.7	nil
Jan–May 87	8	4	180	449	7.4	nil
KAALIN						
Mar–Dec 86	6	13	381	417	21	22
Jan–Jun 87	8	9	121	381	6.8	nil

cultivators. In contrast, there was much less change in the programme of work at either Kaalin or along the lake. Interestingly, in the latter case there was an increasing number of requests as it became recognized that the gardens were retaining large amounts of water and had the potential to produce larger yields than crops grown by traditional methods.

CONSTRUCTION TECHNIQUES

It has been suggested by many development workers that rainwater harvesting is relatively straightforward from a technical point of view, and the chief problems it involves arise from the social context and from questions of organization. Whilst the main justification for this book is its discussion of the social and institutional issues, there are significant technical difficulties which have too often been overlooked. Certainly there is very little successful experience of rainwater harvesting involving earthworks in East Africa apart from some microcatchments used for the establishment of trees, and it seems clear that not all the technical difficulties have been thought through. Hogg (1988) argues that previous lack of success in Turkana was due to underestimated labour costs and the

Building bunds. Men, women and children take part in garden construction.

Moving earth for bund building.

Women carrying earth to build rainwater harvesting structures. They use a shallow metal dish, carried on a cloth pad on their heads.

A later stage in the bund building process: the crest of the bund is aligned with the sticks placed earlier.

The Ethiopian ard – the traditional chisel plough. The metal component for which blacksmiths are needed is the single tine, which functions as a plough-share. In Turkana, this is made from steel scrap (or strip) brought in from Lodwar.

Ploughing with the Ethiopian ard and two oxen. Gardens are ploughed before the rains to increase infiltration.

A garden after the rains have flooded it. Approximately 30cm depth of water is sufficient to bring a crop of fast-growing sorghum to maturity.

The sorghum crop ready for harvesting. Sorghum heads are harvested individually as they become ripe. The stover is used for animal fodder.

high degree of organization required, but in an almost chance remark, he mentions schemes which fell into disrepair as a result of inadequate control of water leading to bunds being breached. Such problems were experienced in many TRP schemes.

One reason why this specifically technical kind of failure is so common is indicated by Finkel (1985) in his manual on water harvesting, where he makes the point that in Turkana District there is inadequate information about runoff flows, flood discharges, and flood duration. In the absence of such data, there has been a tendency to underestimate runoff. This, coupled with a lack of technically competent supervision, is the major reason for the disappointing record of TRP, which had 90 per cent of its earthworks fail in the first two years (even though many helped produce harvests in their first season, making 1982 a particularly good year for sorghum production; Martin and Gibbon, 1987, p.40).

To address these problems, Finkel argued the case for a standard design which TRP staff could be trained to build and then continue to develop after basic training. This raises questions about the most appropriate way of initiating people into a new technique as well as raising more strictly technical questions as to how one arrives at a standard design capable of being adjusted to meet the varying conditions found in the district.

One design criterion relevant to this is the ratio between the area of the catchment providing runoff for a garden and the cultivated area in that garden. Finkel noted that accepted rules of thumb for this ratio vary from 5:1 to 40:1 depending on rainfall, the catchment surface and the crop being grown. Based on knowledge of how much water different crops require together with estimates for rainfall, Finkel calculated that catchment/ cultivated area ratios at Kaalin and Lokitaung should be 23:1 for sorghum and legumes, but only 8:1 if rainwater harvesting were used to support grasses and trees. By contrast, he thought that the appropriate ratio at Loarengak, where rainfall is less, should be 37:1 for sorghum and 11:1 for trees and grass (Finkel 1985, p.10).

In calculating these figures, Finkel had to make assumptions about what percentage of rainfall during a storm runs off across the ground surface to join flood flows in water courses. This percentage is the 'runoff coefficient', and there had been no reliable measurements of it in northern Turkana. To make his estimates, therefore, Finkel was forced to use data from Baringo, a part of Kenya well to the south, with more vegetation cover and different soils. Moreover, the annual average figure which Finkel used masks differences between runoff in storms with different rainfall intensities, and a 'typical' storm in Turkana is unlikely to have the same characteristics as a 'typical' storm in Baringo.

Although he was aware of these limitations, Finkel nevertheless went ahead and used a runoff coefficient of 18 per cent to arrive at his suggested criteria for catchment/cultivated area ratios. Unfortunately, though, experience has shown that this is a wholly inadequate figure, bearing no resemblance whatsoever to the voluminous runoff flows which occur periodically in Turkana storms. The largest storms are violent and intense. Data collected in 1980–81 at Lokitaung suggest that a runoff coefficient of 40 per cent or more is common, and in March 1981, one downpour yielded a massive 64 per cent runoff. More recently, in late 1989, an even more spectacular storm occurred when runoff may have exceeded 80 per cent of rainfall. Not surprisingly then, structures designed on the basis of the 18 per cent figure receive much more runoff water than they can retain, and their spillways have to pass very large flows. Design therefore needs to balance the need for an adequate catchment for collecting runoff in dry years against the threat of large storms, yielding large amounts.

An additional feature of runoff events in Turkana is that flood flows can change the alignment of water courses (Hogg, 1988). When the rains begin, because vegetation has died back during the dry season, the soil is very susceptible to erosion and new gullies can form very quickly. Thus badly sited structures can suddenly find themselves in the middle of a large water course, as the project has learned to its cost. It is interesting to note that local people have a name for the runoff which results from intense and violent downpours. They refer to it as 'red water', and by comparison, more gentle rainfall gives runoff known as 'brown water'. Gardeners say that they seldom plant after 'red water' runoff as crops fail to germinate, or grow very badly.

Aware that he understood little about rainfall, runoff, soils and crop production in the area, Adrian decided to avoid using imported design standards and aimed to make greater use of local knowledge. Local gardeners were therefore the effective designers when the Demonstration Project began. Knowledge was pooled and design criteria developed. The team was also able to benefit from the experience of both the Salvation Army project and TRP.

One advantage of the approach developed was flexibility, and design methods were refined and improved in response to lessons learned. It was found, for example, that around Kaalin, where the catchments are smaller and more dissected, relatively small structures can retain water adequately. Along the lakeside, however, catchments are large and flows of water can be considerable. Here it was necessary to adopt more cautious methods and more robust standards. Following each rainstorm, damage was evaluated, and the lessons learned became incorporated in design specifications. For example, it was learned by observation that it is better to use lateral spillways which deflect water round the ends of the bunds (Figure 15) rather than 'through spillways', which tend to concentrate erosive forces in the length of the bund.

The problem of matching gardens to catchments has also been dealt with in this way; avoiding problems caused by lack of runoff measurements, and

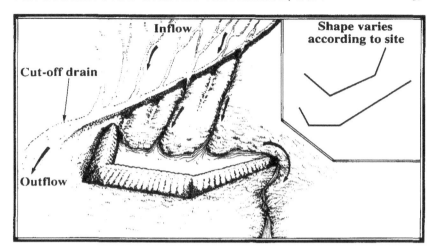

Figure 15   *A bund on an approximately 'trapezoidal' plan enclosing a garden, showing how a cut-off drain can be used to prevent excessive inundation of the garden, with its associated risk of damage to bund and spillway. The diagram also illustrates the principle of the lateral spillway (on the right).*

making catchment/cultivation area ratios seem rather academic. Staff now study catchments closely to establish where the main drainage lines run and how big a channel has been eroded where they converge. The dimensions of this channel give a measure of the volume of runoff flows which is then used in designing earthworks.

Although staff were taught to think on their feet and learn by observation rather than following set rules, they did eventually arrive at a set of guidelines. They were laid down by the team and are still being developed. They mention such things as avoiding areas of clay soils which make unstable bunds, avoiding proximity to large water courses, and avoiding sites which would require extensive levelling. If a site seems satisfactory the procedure is that the team member will sketch it, and then go home to think about the best way to design the garden. The guidelines say, 'sleep on it at least once ... make a design sketch and return to the site to think through the implications. Go with another designer if possible'.

The sketch will normally show a cut-off drain (Figure 15) to protect the garden against excess flow. It should always show the alignments and heights of bunds and the positions of spillways. Once all this is established, the designer will peg out the site with sticks which clearly indicate the final garden layout. Heights of bunds are marked on the sticks, and the opportunity is taken to discuss details with the garden's owner.

When construction begins, the designer will return to the site to ensure that the cut-off drain is complete before work is started on the garden proper; and then will later return again to check that bund compaction and land levelling work are being carried out to an adequate standard. The designer will usually seek specialist assistance in siting the lateral spillway so as to retain only 25–30cm depth of flood water in the garden. The tips of the spillway are protected with stone walling if stones are available. Then finally, there is a visit to check that all the work has been completed satisfactorily before the last instalment of maize is given to the owners.

Adequate compacting of the bunds is one difficult aspect of the work

Figure 16 *Using animal draught power to build a bund. The operator can arrange for the ox-scoop to pick up soil, or simply to transport it like a sledge, according to the angle at which he (or often she) holds the handles. When the scoop reaches the point where the soil is to be deposited, the handles are pushed up towards a vertical position from where the continued pulling of the animals turns it over. The trampling of the oxen helps compact soil in the bund.*

because it is impossible to convince people that this is worth the extra effort. One of the best ways to achieve good compaction is to use animal draught power, because as the animals move soil to make a bund, they are constantly crossing it and thereby compacting it (Figure 16). However, as Adrian notes,

> It would be a mistake to suggest that all technical problems in Turkana rainwater harvesting have been solved. Getting a balance between enough water to flood the sites in a year of poor rains but not so much at other times that they are swept away, is an insoluble dilemma. Sometimes one feels nearer the optimum and certainly there are fewer bund breaches. But it is rarely possible to be sufficiently vigilant and maintain high enough standards. There is also an element of bad luck, like the one-in-twenty year flood, or a river which changes course in a way which was impossible to predict.

To reduce risks, smaller rather than larger catchments are now developed. Attempts are made to reduce the size of catchments by dividing them and deflecting water away from a garden system. Furthermore, gardens are increasingly built with cut-off ditches that will enable them to be isolated from the catchment, if required. A small 'gate' can be quickly closed by shovelling earth into the opening, closing all inflows.

MAINTENANCE PROBLEMS AND ANIMAL DRAUGHT

Since it is inevitable that bunds will sometimes be breached and some repairs will be necessary after heavy storms, it was important to assess the preparedness of local people to carry out such work. TRP had paid for repairs through food-for-work but it was eventually decided to make garden owners entirely responsible. The rains of April 1987 were the first major test of this policy. Two bunds had been breached because of design faults. As staff members were to blame for these, the project paid out maize for their repair. Other breaches were due to 'piping' under bunds. This phenomenon occurs most commonly in soils with a high clay fraction and is due to the shrinking and cracking of clay masses as they dry out. When water builds up behind a bund which is in this condition, it may begin to percolate through the cracks, gradually eroding a wider 'pipe' for itself. The policy now is to avoid working with this type of soil and to warn gardeners when staff visit sites and see cracks beginning to open.

Strangely, it was the experience of repairing large breaches in bunds that led to a more meaningful use of animal-drawn equipment. As several of the gardeners complained about the amount of work which needed to be done, the *ekarabon* (team leader) at Kachoda encouraged them to use a team of draught oxen. This they did, agreeing to water and graze the animals, and a new dimension in the work opened up.

Because progress in securing acceptance for animal draught had been slow, Adrian had appointed a down-country Kenyan, Simon Barasa, as animal draught co-ordinator. He had first worked in the District with Brian Hartley at the spate irrigation scheme at Loarengak. After a period of induction, Simon's first task was to develop a range of animal-drawn equipment which could be manufactured locally. This was hampered by the fact that there were no skilled metal workers in the area. The Turkana have historically depended on the Labwor people of Uganda for their iron tools and weapons, and have only developed limited skills such as the beating of cold metal to shape it. When on local leave in Marsabit District, Adrian sought out some local Boran blacksmiths and arranged that they later travel to Turkana to train local staff, hoping that in this way, staff would be able to develop a basic *ard* (plough) and a range of hand tools themselves. Unfortunately, the plan did not work at all well, and the blacksmiths were soon involved in establishing a local trade, and not training the staff. Things went from bad to worse, and within a short time the project was forced to sack them. However some *ards* were made before this happened.

Eventually (and beyond the period covered by this chapter), Simon resolved the difficulty by employing two apprentices (one a school leaver). Then later, he persuaded a Samburu smith, Elijah Lolokoru, to spend a month with the project, training the team and making some Ethiopian *ards* (See photo section). Elijah came during September in 1989 and the standard of his training was extremely high. The trainees were each ritually initiated into the 'blacksmith's clan', which cuts across ethnic divisions. Blacksmiths are a revered group in many parts of Africa and are thought to have special powers.

In addition to developing the equipment, Simon assisted field staff organize simple field trials to determine the daily work rates of oxen and donkeys. Girth measurements were also taken, making staff more aware when animals were losing weight. From the trials it was learned that a single pair of oxen is more efficient than a team of four, despite the fact that a four can carry more soil per load. For example, in trials at Lomareng in May 1986, carried out over a ten-day period, one pair of oxen ploughed an average of 954m^2 per hour, whereas a team of four oxen, working as two pairs, averaged 1436m^2 per hour. Thus using oxen in pairs was 25 per cent more efficient than using the larger team. Pairs are easier to handle, in particular to turn (see photo section), and less time is lost during working hours. In most cases, soils are light and easily opened, so it was decided to work only with pairs.

The trials were also useful in that they helped project staff establish a code of practice for working the livestock. The animals were fed only local fodder and thus it was all the more important for adequate time to be allowed for grazing. Thus a standard working day of 4.5 hours for oxen and 2.5 hours for donkeys was established when fodder supplies were adequate.

Draught animals were then worked shorter hours, adjusted according to their weight loss, when fodder was short. If weight losses were persistent, the animals were sent to join the breeding herds in the dry season grazing ground.

Despite these technological developments, it was disappointing that people left all the decision-making in the hands of the project. No matter what was tried, people did not want to become involved, nor to allow their own animals to be used. Staff were about to give up completely when the damage to bunds following the 1987 rains led to a change in attitudes. Several gardeners were faced with the choice of either making extensive repairs or abandoning their gardens. Without any form of food aid, their only incentive was the anticipation of crops available to the household. This being the case, families were left to make their own decisions, and were given no support. Several requested maize, but this was turned down. However, a member of staff chanced to remark that gardeners should at least be allowed to use draught animals for repair work. This was discussed thoroughly, and it was agreed that the animals could be loaned, provided that the family worked, grazed and watered them rather than having members of staff to do this.

During the season, several gardeners borrowed animals on these terms, and it was agreed that in future years the same system should operate. In order for each local committee to be able to offer the service, it was also agreed to divide the draught animals between the committees at Kachoda, Kaalin and Loarengak. The committees then took full responsibility for the animals throughout the year. A change in 'ownership' was thereby effected. This system has worked reasonably well and there have been no more cases of abuse of animals.

HARVEST OUTCOMES

In 1986, the Dutch and Kenyan governments commissioned a study of the economics of rainwater harvesting in the arid and semi-arid lands of Kenya. This was carried out by Richard Hogg, and came after a period when enormous quantities of maize had been distributed to pay for bund construction in Turkana. When the value of this maize was calculated, the cost of gardens made under the auspices of the Demonstration Project appeared to be very high – over US$300 per hectare. But this figure did not allow for the way in which the project was steadily reducing costs. As indicated in previous paragraphs, people were now willing to make gardens without large food-for-work payments because they believed the work to be worthwhile, and gardens were now being designed so that less levelling was necessary, and less elaborate spillways. Reduction in volumes of earth moved and food distributed is indicated in Table 7. The result was that although food-for-work costs in the first year, 1985, were the equivalent of

$417 per hectare, they fell to $254 in 1986 and $136 in 1987. Even so, rainwater harvesting works are substantial investments of money and effort. To justify them, it is necessary to demonstrate that there are benefits and that local people are able to maintain production over time. It is also necessary to look more widely at the impact of improved rainwater harvesting on the household economy.

During 1985, there were only four gardens completed in time for the rains. Some crops were established but only two gardens produced a harvest, with yields estimated at about 400kg/ha. Whilst this is a minimal return on investment, it was clear that the gardens were well cared for and that families were prepared to put their labour and time into producing this harvest.

Before the 1986 rains, an additional fourteen gardens had been developed in the Kachoda area, bringing the total to eighteen. All the gardens were planted and good crops were established, but about three weeks before harvest, the District was invaded by a plague of grasshoppers which devastated the fields so severely that one could get the impression that the crops had been grazed off by cattle. Consequently, there was little or no harvest that year, apart from stems which people harvested to eat (in the same way as sugar cane is eaten). Thus in two years there had been little return on investment.

Despite the lack of progress, additional food-for-work was available and more gardens were developed during the dry season of 1986 and early 1987. Nearly fifty were ready for planting at the start of the 1987 rains. While the majority were in the Kachoda valley, the project was now also working from Kaalin (where there were six gardens) and from Loarengak (which had a similar number at two lake-shore sites). Good crops were established in nearly all these gardens and it was very much hoped that this third year of effort would result in useful harvests. However, results were again poor. Most of the gardens around Kachoda had been planted with Merille sorghum, rather than with the local Turkana variety. In a resurgence of traditional bartering with neighbouring ethnic groups, Merille grain had been available at half the price of Turkana sorghum. Most gardeners had therefore bought seed of the Merille type. This reflected the lack of experience of those new to gardening. Lake-shore cultivators, with their detailed knowledge of the different varieties of sorghum, appreciated that Merille sorghum was less likely to yield in a dry year and therefore commonly planted mixed stands of Merille and Turkana sorghum, or just the Turkana variety. Thus gardens in the Loarengak area produced reasonable amounts of grain in 1987, some of which was bartered to obtain goats. At the end of the season, owners of improved gardens had exchanged sorghum for varying numbers of goats, averaging 4.4 per garden (Watson, 1990, p.34). Since about 50kg of grain was required to obtain a goat, this implies that on average the gardens produced over 200kg of sorghum *in addition* to what was eaten by the family or shared with friends.

One particularly interesting garden was at Natoo, 8km east of Lokitaung. In early December, 1986, a localized storm flooded the garden to a depth of more than 25cm. The owners (two related families) planted Turkana sorghum shortly after and in early February, harvested a small crop. In addition to that eaten in the field, more than 120kg was stored for later consumption (although in fact some was traded for a goat).

Shortly after the harvest, there was another rain storm which resulted in runoff and flooding. This stimulated the growth of a ratoon crop and a second small harvest in March. In April, the long rains resulted in further flooding and the garden was replanted, again with Turkana sorghum. A very good crop was established and the harvest was taken in June. Showers in June and July led to the growth of an additional ratoon crop, which was harvested in early August. Thus in under eight months, the families had taken four harvests.

The improvement of this garden had been paid for with twelve bags of maize. It is estimated that, in addition to sorghum eaten in the field, the total produce from the four harvests on this plot was more than 300kg (about 735kg/ha, worth more than five goats to the family). If sorghum eaten in the garden were included, these figures might be 20 per cent higher. Thus in only one year, a quarter of the food aid used to develop the garden was produced. This made it clear that where Turkana sorghum was planted, good results could be achieved even in years of less rainfall. The lessons learned led to the establishment of stores which would sell seed, described in the next chapter. They help to ensure that seed of the Turkana variety is always available.

6. THE START OF SELF-MANAGEMENT, 1987–90

LEADERSHIP TRAINING AND LOCAL CONTROL

By the beginning of 1987, the work on rainwater harvesting described in previous chapters was distributed between three locations, each of which had its own committee. Some meetings of elders from all three areas had also been held and had proved to be useful sessions for improving communications between myself as project manager and local staff. From them grew the idea of an executive committee for the whole project.

At this stage, it was felt that if further progress was to be made, the staff (and Adrian also) needed training in leadership skills. An approach was made to World Neighbors in Nairobi and, after an initial exploratory visit, it was agreed that a leadership training course should be run in February 1987. Based partly on the DELTA technique (Hope and Timmel, 1984), the course would discuss ways of encouraging self-reliance and the capacity to plan and manage projects. Although the staff were at first sceptical, most came away very enthusiastic for what they had learned about the possibilities of self-directed development. The course did much to increase the level of self-confidence, and they felt that another should be held to include representatives from the local committees. Adrian too was challenged by the course, and from this time onwards, felt a clear (if unstated) commitment to carry the project through to the point where local people could take over entirely and manage the project without external support.

The rainwater harvesting work had only been funded for a limited period, towards the end of which a decision had to be taken as to whether and how to continue. In July 1987, recommendations on this were put forward by advisers from Britain (Martin and Gibbon, 1987), and were discussed with the local team leaders (*ekarabon*) and then by local committees. The result was a unanimous decision that people did not wish their rainwater harvesting work to be taken over by TRP. Rather, they would run the project themselves, and apply to Oxfam for any necessary financial support. A meeting to discuss this was held at Kaalin on 23 July 1987, when Adrian went through the advisors' recommendations, discussing them with the elders from the three locations, namely Eyanae (from the lake shore), Ekitela (Kachoda valley) and Akorot (Kaalin). The minutes of this meeting demonstrate a very positive attitude:

Akorot I started work here in Kaalin in '86 during the year when no-one knew anything. In '87, people began looking for the rain and wondering if anything was going to happen. Now everyone is trying to push the work on in their own gardens, even when the *fundi* doesn't go to work.

Eyanae It was good that Oxfam came and built only one garden at first or else people could have been angry . . . because they did not believe it would work. But now everybody is looking for a garden. If I go to a wedding feast, I hear people saying, 'What about my garden?' If I go to a goat feast I will hear them saying, 'What about our gardens?' When I am at home, when I am at work, people come in small groups, in large groups, I only hear all the people of the lake side demanding gardens.

Ekitela Two things remain. We need to think about the manager and the extensionists . . .

Akorot We do not want a TRP manager, neither a manager at all. If it is time for Adrian Cullis to leave, then we will take over . . .

Ekitela We do not know how to write reports and letters. We need a small office where we can collect information, keep the accounts and things. If we had somebody like Meyen (ex-manager of the Salvation Army project), we could easily carry on. What we need is a good *karani* (secretary). But we need to make sure he does not become a manager – otherwise he will grow horns. The committee should be able to handle this person.

Oxfam's country representative in Kenya, Nicky May, was very encouraging about the local committees taking charge without a manager, along the lines discussed here, and suggested a number of points on which the local committees would need to take decisions. Her views greatly assisted the formation of effective administrative procedures and smoothed the transfer of responsibility to the executive committee, now functioning as a Management Board.

Full responsibility for the project was handed over on 30 July 1988, but the Board had actually been running the project for six months before that. The idea had been that there should be a transitional period when the project would be under local management, but Adrian would still have a supporting role. In the event, he was absent for a long period recovering from hepatitis, and this delayed the first stages of the hand-over, and disrupted some training courses. After the Board took over, Oxfam agreed to continue financial support for a three-year period (up to April 1991).

The point to be underlined is that the importance of institutional development was now fully recognized, and the structure which had evolved – local committees and a central Management Board – was about to be tested.

The three elders (*ekarabon*) whose discussion in July 1987 has been

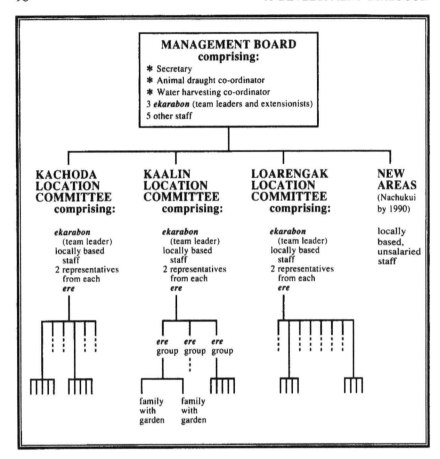

Figure 17   *Organization diagram as at 1988. By 1990, the location committees each had a chairman and chairlady, and the Management Board had been enlarged to include these chairpersons and more staff members. (An asterisk denotes a salary paid by Oxfam.)*

quoted were clear that they wanted the locations to retain their mutual links, but that they also wanted to develop their own programmes. There was also the question they had raised about who would write reports and letters. The answer found for both points was to appoint a Project Secretary, who would be a co-ordinator, responsible to the Management Board for such duties as reporting to Oxfam, paying salaries and expenses, preparing budgets, collecting data, and organizing monitoring and training. There would also be two other co-ordinators dealing with more specialist matters.

BOARD, BUILDINGS AND STAFF

Though it had evolved from an earlier executive committee, the Management Board was in many ways similar to a traditional institution, the *ekitoi a ngikasakou*, or elders' meeting tree, which is found wherever groups of Turkana are living. A shady tree, commonly in a river bed, becomes a place where elders meet daily to share news and discuss grazing strategies. Membership of the board consisted of elders (*ekarabon*), who functioned as team leaders and extensionists, and was strengthened by participation of senior administrative and technical staff. All three locations were represented by *ekarabon* and other staff (Figure 17).

In many ways, the Management Board structure mirrored that of local committees, each of which could also be compared to a traditional elders' meeting except that all members were salaried. Now, however, representatives of locations who are not salaried are more active on the Board. Despite initial inexperience, Board members soon proved their resolution and commitment. At their first meeting after hand-over in July 1988 a senior *ekarabon* (Akorot) was retired in favour of a younger, more active elder. It appeared that a number of irregularities and disputes had been handled badly at Kaalin, and as a result, the Board thought a more energetic person should be leading the local team.

Another decision taken by the Board was to appoint Pius Chuchu as Project Secretary. This was against Adrian's advice, for indeed he had worked long and hard to convince staff that another candidate would be more suitable. He did not insist on his view, and the *ekarabon* had the last word. Time and again, he had to re-learn this lesson, that it is better to go with local respected elders rather than driving home one's own ideas. Pius has proved to be an ideal Project Secretary, and there was a significant parallel between his appointment and the replacement of the *ekarabon* at Kaalin. The new person in this job, a man named Losiru, was the first elder to be appointed by the Board, and it was interesting that with both appointments, members seemed to stress the recruitment of people who were well-known, trusted, gentle, and able to work in a team. Selection criteria favour character traits over 'qualifications'; Pius Chuchu only had primary education yet was selected against a field of better educated applicants.

One function of the staff – particularly the co-ordinators – was to record data relevant to monitoring progress. It seemed desirable before self-management was introduced to simplify record-keeping arrangements in order to reduce the work load and improve the quality and flow of information for planning. For example, it was agreed that the Project Secretary should keep monthly accounts, minutes of Board meetings, and be responsible for the quarterly reports to Oxfam and to the Ministry of Agriculture in Lodwar. The co-ordinator for rainwater harvesting, by contrast, is responsible for checking that rainfall figures are collected in each location. He is also responsible for food-for-work contracts, store cards, and records of floodings, breakages of bunds, repairs, and crop yields.

The transition to self-management involved some consideration of buildings as well as of staff duties. During its first three years, the project had been managed from Adrian's house, and then from a rented shop in Lokitaung. It was now felt that Oxfam's agreement to fund for another three years justified the establishment of a more satisfactory office and store. The opportunity was taken to move the centre to Kaalin, where it was hoped that staff could focus more specifically on their work, rather than being caught up in local politics and in dealing with the administration in Lokitaung. The building was designed and constructed by the staff themselves. It was made of local materials where possible – stone and mud, with some cement rendering – but with a 'tin' roof.

As the transition to self-management approached, concern was expressed in local committees about loss of salaried staff, longer-term funding, and transport. Questions were also raised about the extent of Oxfam's commitment. The latter appeared to reflect local perceptions of development workers who quickly move into and out of projects. Time had to be spent reassuring local committees, and through them, the *ere* members.

One way that Oxfam's longer term commitment could be demonstrated was to provide tangible support for programme development. There was growing interest in the establishment of local stores where grain, seeds, and skins could be held on behalf of members of garden groups. In July 1987, the Kaalin local committee was already building such a store on their own initiative, though at Kachoda and Loarengak, the possibility was only just starting to be discussed. At this point, Oxfam agreed to pay for roofing materials for any of the committees which had built the walls of a store.

It was thought that local committees would benefit not only from improved food security and better availability of seed for the correct variety of sorghum, but also because the stores would provide a focus for group activity. Visitors and government officers would be more easily able to locate staff, which would improve relations and reduce frustration. The stores were built by the local members using voluntary labour with support from staff, particularly from the rainwater harvesting technicians who were skilled in stonework. Gardeners' *eboka* payments were used for funds, and the Oxfam roofing grant paid for metal sheet, cement, and weighing scales (so that food could be sold from the stores).

Although local committees worked in much the same way as traditional elders' meetings, one change encouraged by the project was a greater representation of women. Whilst steady progress had been made to persuade women *ere* representatives to attend the monthly meetings of the local committees, it was found more difficult to gain acceptance for the representation of women on the Management Board. However, if a woman had specific skills and a definite role in the project based on use of those skills, there was ultimately little objection to her membership of the Board. Moreover, qualifications in a particular area of expertise helped give some of the women more confidence in participating at meetings. Thus the acceptance of women in decision-making was linked to their access to training.

At the time of hand-over to local management in July 1988 the project employed 16 people, three with responsibilities covering the whole of the work and thirteen based in the locations (Figure 17). These included only five women. The male employees included two monitors seconded to the project from TRP. Later in 1988, their contracts came up for review, and they were replaced by two women who had proved their abilities by helping with a separate research project on the development needs of Turkana women (Watson, 1988). Not only did this help improve the gender balance among the salaried staff, but it resulted in women being appointed to the Management Board, whose eleven members now included four women.

Subsequent to the building of stores at each location, additional staff were recruited (or transferred from other work, notably as monitors) to manage food sales and the growing trade in skins. In two places, women were given these new roles. A further development was that each local committee elected both a chairman and a 'chairlady' (Martin, 1990). These people count as unsalaried staff, and are members of the Management Board. This and other changes, including work at a new location (Nachukui) has meant that by mid-1990, the salaried staff numbered seventeen, of whom five were women. At the same time, the Management Board had increased through the addition of chairpersons from the three locations, and now had a membership of eighteen, of whom six were women.

EQUITY AND TRAINING

Project policy toward women can be said to have several components, of which representation on committees and appointments to staff represent only one. Training policies are linked to this since it is through acquisition of skills that women become eligible for staff membership and accepted on committees. The test of these policies, of course, is whether they lead to the project becoming more responsive to the needs of women in the pastoral community. There is evidence of this in the help given in developing gardens for women-headed households. The stores also are very much a response to problems identified by women. However, Martin (1990) has noted a 'lack of discussion of women's issues and interests' on the persistently male-dominated Board, and has suggested more open discussion of gender issues on training courses.

The latter would particularly refer to leadership training and awareness raising. Dealing with these themes, the World Neighbors team carried out a programme of three workshops in which staff members were invited to participate. In August 1988, an effort was made to offer similar training to selected *ere* leaders from the locations. Unfortunately, though, the TRP extensionists responsible failed to give the type and level of training required, and it was decided to abandon the venture. Following this disappointment, staff asked for a follow-up to the World Neighbors workshops designed to help them lead training sessions themselves. Oxfam's pastoral projects officer, Peter Kisopia, agreed to co-ordinate these workshops. Much more has yet to be done, however, to develop training material which is culturally sensitive, and which would enable courses to reinforce the nomadic pastoralist culture. Most of what is available is aimed at 'settled' populations.

It will be recalled that, apart from the salaried staff, the project had been training many members of *ere* groups in rainwater harvesting techniques, for example in the use of line levels (Chapter 4), and animal draught (Chapter 5). Many of those who received this training played a leading part in garden improvement and rainwater harvesting work, helping their neighbours in the same home area or *ere*, so that they are sometimes referred to as '*ere* technicians'.

By 1987, the emphasis in technical training had slowly shifted away from the more formal methods described in Chapter 4. Instead of training courses, there were now training days, and an apprentice style of training was also operating. 'Apprentice training' meant that trainees worked daily with staff, which was felt to be a more sensitive approach culturally, enabling considerably more 'informal' information to be handed on.

These arrangements worked best while food-for-work maize was available, because *ere*-level technicians received a proportion of the ration. In return, the technicians would support and guide gardeners whilst bunds were built and levelling carried out. However, when reliable maize supplies ceased, it has been more difficult to motivate *ere* technicians. Some continue to do excellent work but where others have drifted away, the work of salaried staff has become more difficult, as they are now left to provide all the technical support to gardeners.

Once less reliance could be placed on trained individuals within each *ere* to help gardeners, it was logical to consider how more skills could be transferred to all the gardeners. To take an example, when the animal-draught workers on the staff are asked to assist a garden owner, they will take the oxen to the garden and start the cultivations. As the morning continues, the gardener will be asked to help and then to hold the plough. Next day, the gardener will assist in yoking and working the animals. In the next season, the staff member may only visit the garden on the first day, expecting the gardener to remember the previous year's experience, and to learn more quickly. Increasingly, then, the onus is placed on the garden owner.

Looking beyond these activities, staff expressed an interest in visiting other similar projects. With Oxfam support, a number of joint workshops for Oxfam pastoral project personnel were organized. One was held at the Samburu Rural Development Centre, Maralal, and a team of six travelled from Lokitaung to take part. In addition to discussion sessions, participants visited restocked communities and group ranches. Another workshop was organized at Loitokitok, Maasailand, in February 1988. This visit was particularly interesting as for the first time, staff had the opportunity to witness cattle herding managed under 'group' ranches.

One encouraging development is that exchange visits and the workshops associated with them are now being used as means of generating ideas for developing Oxfam's pastoral policy in such a way that the ideas and suggestions of external consultants need less often be sought. Another innovation has been the appointment of a pastoral projects officer, Peter Kisopia, who is himself a Maasai.

SOIL, WATER AND CROP MANAGEMENT

At the end of 1987, when food-for-work payments were cut throughout the District, work on TRP water harvesting sites almost stopped, and it became clear that many gardens were being abandoned. However, the Oxfam project gardens were less affected, though the rate of construction slowed and design changes were made to reduce the amount of earthmoving. At the same time, people on the lake shore maintained their interest in improving their gardens with help from the project.

Awareness of this within other agencies and in the Ministry of Agriculture meant that the project received increasing recognition and interest. In February 1988, for example, Adrian was invited to lead a session on extension methods at a water harvesting workshop at Lodwar arranged by TRP and the Ministry of Agriculture. Whilst the discussion there was interesting, it was difficult to discern any real commitment to small-scale rainwater harvesting along the project's approach.

While some food-for-work was still available, contracts were simplified so that payment was at a flat rate rather than dependent on the volume of earth moved. The rate was set at six bags of maize for the household, and one for the *ere* technician, and households were still expected to contribute an *eboka* payment consisting of one goat. Despite this, the numbers of gardens being developed continued to grow, and many bunds were well made and nicely finished. In several cases, gardeners on the lake shore actually built larger bunds and better spillways than specified in order to increase safety margins. Moreover, the introduction of the flat-rate contract enabled the entire process of garden improvement to be handled by the local committees, since it was no longer necessary to estimate volumes of earth being moved. Staff could very competently manage everything else, including site selection, garden design, construction and levelling.

Some *ere* groups on the lake shore chose to tackle formidable levelling operations. In one garden, a gully fifteen metres long and one metre wide was filled in. The provision of wheelbarrows considerably helped here. Many were rescued from TRP and repaired.

Having learned from the 1987 season, when a shortage of local seed resulted in the planting of Merille sorghum, it was agreed to establish revolving seed funds in each location. More than 385kg of local sorghum seed were purchased from TRP and distributed to the local committees. The sorghum was offered for sale to members and other gardeners at KSh5/- per kg and the proceeds were used to establish a revolving fund. The provision of seed encouraged a number of gardeners to dry-plant their gardens rather than waiting for the first rains.

It is worth describing the performance of the gardens in the 1988 season in some detail, since this was the first year of self-management and many of the practical problems experienced previously had been overcome. At Kachoda, (for which data from monitoring at five *ere* are presented in the first column in Table 8), all gardeners purchased their seed at the store. The rains started in April and there were three storms producing large amounts of runoff into gardens. As catchment size and inflow arrangements vary, the depth of flooding was often quite different between one garden and another in the same area. There was little damage to the earthen structures at any site due to excess flooding and none where cut-off drains had been installed.

A total of fifty-one gardens were planted before the end of April. Some people dry-planted before the rains and a little replanting was necessary. To avoid two gardens remaining unused because herders were away with their stock, one member of staff rented the land and organized a 'beer brewing' for members of her *ere* in payment for their help in planting it.

The production figures shown in Table 8 were measured by the owners using 1kg *Kimbo* tins. The result was a reasonably accurate record of threshed grain, but additional sorghum was eaten in the fields, so the data tend to underestimate yields. Ways in which sample households disposed of their sorghum were also recorded and are noted in the table. Almost a quarter of the sorghum was given away, in most cases to reinforce social networks (whose importance was stressed in Chapter 3).

The local store at Kachoda was completed before the sorghum crop was harvested and nine gardeners used it to store a total of 544kg (almost 7 sacks), from which withdrawals were made over a period of 2-3 months.

Table 8 Results of monitoring gardens during the 1988 growing season. (Source: Cullis, 'Handing-over report' December 1988; also Martin, 1990)

	KACHODA	KAALIN	LOARENGAK
Total number of gardens with bunds completed before rains	69	27	31
Total number of gardens planted	51	23	30
(dry-planted before rain)	(15)	–	(5)
Dates of storms producing full or partial flooding at some gardens	6 April*	27 March	10 April
	10 April	25 April*	18 April
	26 April	26 April	20 April*
	30 April	28 April	26 April
	12 June	30 April	30 April
Numbers of gardens producing:			
a) significant harvests	48	17	30
b) harvest failure	3	6+	1
Sorghum production recorded on *monitored* plots:			
a) largest recorded	397kg	60kg	824kg
	(1323kg/ha)	(200kg/ha)	(1268kg/ha)
b) smallest recorded	25kg	9kg	222kg
Average production in *all harvested* gardens	52kg	50kg	228kg
Total grain produced in all gardens	2.48 tonnes	0.9 tonnes	6.87 tonnes
Other crops	cowpeas maize greengram water melons	none recorded; figures approximate	cowpeas
Use made of sorghum grown (sample of 5 *ere* at each location):			
a) given away	23%	data	29%
b) bartered	9% (for goats)	incomplete	35% (for sheep & goats)
c) eaten at home	40%		24%
d) stored (at time of survey)	28%		12%

* *full* flooding on at least two sites;
+ some crops destroyed by waterlogging.

At Kaalin, monitored gardens were all fully flooded at least once during the growing season. No damage was done to bunds or spillways, but gardeners reported that when their plots were flooded a second time, sorghum plants were killed by waterlogging. Crops were established in seventeen gardens altogether, although some produced very poor harvests. Indeed, the Kaalin committee was not very active during the season and little effort was put into weeding and bird-scaring. Yields reflected this lack of enthusiasm.

At Loarengak, as in Kachoda, a number of gardens were planted before the rains, and in some cases, this strategy was successful. In others, where the cut-off drains were inadequate to prevent a second flooding and plants were killed, gardeners had to replant. The rains began in April, and during the month, 156mm were recorded in Loarengak. After that, no rain fell during May and June. Floods did little damage to bunds, but all the sites recorded extensive damage to cutoff drains. Many of these have had to be completely re-dug and others realigned. In some gardens, flood water took up to eight days to infiltrate completely. As a result, planting operations were extended into the first few days of May, by which time the crops in Kachoda were already well established. Gardeners were active in keeping weed growth to a minimum and good crops were established in thirty of the thirty-one gardens along the lake shore. Production data were collected from all thirty plots, so it is known fairly precisely that sorghum harvested totalled 6866kg. Output per garden varied between 10 and 896kg. These figures again underestimate yields because of grain eaten in the fields and occasional sorghum heads cut before harvest and given to friends. Over a quarter of the grain was given away, as at Kachoda, and about a third was bartered, enabling the five monitored households to add ten sheep and five goats to their herds.

Progress made along the lake shore evoked comment in several quarters. In particular, the District Officer remarked at a meeting in Loarengak that one nearby garden site reminded him of maize fields in Central Province. Such recognition has been very encouraging to staff and committees alike. However, the 1988 season was remarkable in many ways. Good rains extended until September, and in December, cattle were still grazing throughout the Kachoda valley (which is usually only good for wet season grazing). All livestock types were in good condition, and with numbers increasing, Turkana pastoralism was stronger than since 1978.

One anomalous development, however, was that precious stones were found at Nadunga, on the Lokitaung to Lodwar road. A settlement sprang up and large numbers of people moved there to search for the stones, which lay exposed on the surface. Many people left the project area, so that some gardens around Kaalin and Kachoda were abandoned. This emphasized the contrast between these two localities where gardening was not traditionally practised, and the lake shore where experienced cultivators were very keen to improve their gardens. It was clear that families to be helped in the Kaalin and Kachoda areas needed careful assessment to check how interested they were.

RAINWATER HARVESTING IN 1989 AND 1990

Some gardens in the Kachoda valley were made in 1985 or 1986 when food-for-work rates were generous, and were abandoned when the original owners moved away with their herds, leaving nobody from their families to plant a crop. One objective of work done in 1988–9 was to allocate these gardens to new owners and rehabilitate them. Some new gardens were also opened in the Kachoda and Kaalin areas, but garden improvement was more extensive on the lake shore. In all three locations, many of the gardens were ploughed prior to the 1989 rains (which came early) using draught animals. The three committees between them owned seventeen oxen, fourteen donkeys and three camels, but one encouraging develop-ment was that some gardeners were now using their own donkeys for cultivations (Cullis, July 1989). Later in 1989, with help from an outside consultant based in Darfur, Sudan, an *ard* with a modified draught angle was produced, which is easier to work and can be pulled by a single donkey. Design and manufacture of a hand-held weeding tool was also begun under the supervision of Simon Barasa, the animal draught team leader. By late 1989 he was assisted by two apprentices whose training experience with a skilled blacksmith was mentioned previously. This greatly improved the project's capabilities, but the forge and equip-

Figure 18   *Blacksmiths' equipment and methods. (Source: Oxfam video)*

ment used were of a simple, mainly traditional kind (Figure 18) to ensure that the project did not become dependent on specialized parts and equipment from outside the area.

During 1988, the project was able to obtain some hand tools from TRP, together with a supply of broken wheelbarrows, which were repaired and used intensively for earth-moving, especially by men. Gardeners also had better access to dam-scoops for use with ox and donkey teams, and the women were able to obtain better containers (*karais*) for head-loading soil. After rains in February, 1989, crops were planted and as they began to grow, the new weeding tools were distributed to some of the women for trial. It was noticeable that weed control was better than in previous years, particularly in Kaalin and Kachoda, and the tools are now sold through a revolving loan fund.

One major problem, particularly at Kaalin, was an infestation of the sorghum by stem-bore insects. This problem, exacerbated by associated with mono-culture, was a particular problem at Kaalin as people had no knowledge of traditional pest control methods. A small amount of insecticide was purchased for experimental use, but it was decided that this was not something to encourage. Once again, Kaalin yield data were incomplete, but it appears that Kaalin gardens may have produced less than a tonne. By contrast, 45 gardens around Kachoda gave 8 tonnes at the main harvest with a ratoon crop amounting to perhaps 2 tonnes, while 63 gardens in the Loarengak area produced a staggering 15.5 tonnes of grain, a figure which includes some crops grown with runoff from the 'short rains' later in the year. Here and at Kachoda, cowpeas were grown as well as sorghum, and a range of other vegetables was planted at Kachoda, including melons, gourds, and newly domesticated 'wild' vegetables (Cullis, 1989).

After two wet years, rainfall in 1990 was again below average. In the Loarengak shore area, grain production was significantly reduced, with 5 tonnes coming from 76 gardens. By contrast, production in the Kachoda valley was initially much the same as in 1989, though with yield per garden and per hectare a little reduced. But then a ratoon crop contributed an extra 2.3 tonnes, taking overall output from the valley over the 10-tonne level (Martin, 1990, p.13). On the lake shore, there were six gardens at a new location, Nachukwi, from which no yield figures are quoted, but at Kaalin production was better than in previous years. Presumably gardeners were now more experienced, and pests were better controlled, and these factors more than compensated for the lower rainfall.

Detailed figures presented in Tables 9 and 10 summarize the whole of the project's rainwater harvesting results up to September 1990. Table 9 deals with the cost of garden improvements, an issue previously raised in Chapter 5. It is clear that the cost of earth-moving and construction work at gardens has been considerably reduced. Moreover, with TRP failing to deliver some grain during 1989–90, some figures represent food-for-work

Table 9   Food-for-work costs of constructing bunds and levelling gardens. (Source: data compiled by Adrian Cullis)

YEAR	Cumulative total of gardens at year end	Gardens improved during year	Bags of maize distributed as food-for-work	Total value of maize* (KSh)	Average cost per garden	
					(KSh)	(US dollars+)
1985	10	10	365	146 000	14600	417
1986	44	34	758	303 200	8917	254
1987	117	73	875	350 000	4794	136
1988	166	49	343	137 200	2800	80
1989	181	15	105	42 000	2800	80
1990 to mid-year	196	16	112**	44 800	2800	80

* based on Lokitaung price for maize in 1990 of KSh400/– for 90kg bag.
\+ based on 1990 exchange rate of KSh35/– for US$1.
** estimates of amounts due; maize distribution was in fact stopped.

payments owed to gardeners, not payments actually made. Thus a good deal of the most recent work has in fact been done without pay. When food-for-work was reduced to a flat rate in 1988, the pace of garden construction fell (though not as sharply as on TRP sites). The interruption of food supplies in 1989–90 also discouraged some gardeners, though many were keen to push ahead with improvements in any case.

Food-aid costs of $136 per garden in 1987 amount to $400–500 per hectare. By comparison, the cost of some trapezoidal bunds built near Kakuma was equivalent to $800 per hectare, and the costs of the spate irrigation systems at Karubangorok and Loarengak was $625 and $1000 per hectare (Hogg, 1988). An even more striking comparison is the cost of conventional irrigation schemes in Turkana District, stated to be $61 240 per hectare (Hogg, 1988).

In addition to food-aid costs, it is necessary to point out that there are other expenditures associated with garden improvements, including purchases of tools (or materials to make them), and salaries of the design team. Salaries amounted to KSh920/– per garden in 1987, equivalent to $26, but were five times greater in 1989, partly because salary costs were spread over fewer gardens. This is one reason why numbers of salaried staff have been reduced.

There are some other costs which it would be misleading to allocate directly against gardens. For example, Adrian's own salary and travel costs were a short-term expenditure, and gardens are now being developed without his help. More difficult are costs associated with the use of animal draught. Whilst animal-drawn implements are increasingly used, especially

Table 10   Sorghum harvests on rainwater harvesting sites

YEAR	Cumulative total of gardens improved	Number of gardens planted	Number producing a harvest	Total grain output (tonnes)	Typical yield (and comment)
1985					
before rains	4[e]	4[e]	2[e]	0.12[e]	400kg/ha[e]
end year	10[a]				
1986					
before rains	18[e]	18[e]	n.d.	very	small due to
end year	44[a]			small	grasshopper devastatation
1987					
before rains	50[e]	43[e]	10[e]	2.0[f]	Merille sorghum
end year	117[a]			mostly Loarengak	planted in Kachoda; pest damage
1988					
before rains	127[d]	104[d]	95[d]	10[b]	200kg/ha[b]
end year	166[a]			of which 7[c] tonnes from Loarengak	in Kachoda; 425kg/ha[b] Loarengak
1989					
before rains	181[c]	131[c]	n.d.	26[c]	1000kg/ha[b]
year end	181[a]			of which 15.5[c] from Loarengak	exceeded around Loarengak
1990					
before rains	n.d.	146[b]	n.d.	15[b]	–
mid-year	196[a]				
September	215[b]				

Sources: a – Table 9 above (data from A. Cullis)
b – Martin 1990
c – Cullis, July 1989
d – Table 8 above (data from Cullis, December 1988)
e – Chapter 5, text above; data from A. Cullis
f – calculated as 10 gardens producing an average of 200kg each.
n.d. = no data

for bund repairs and tillage, most gardens continue to be built without animal draught.

It is clear from all this that rainwater harvesting is a relatively costly measure, and it is therefore necessary to demonstrate that there are benefits in terms of crop production. Thus Table 10 brings together figures already given in the text to show that in the three most recent seasons (1988–90), considerable amounts of sorghum have been produced. Since rainfall was exceptionally good in 1988 and 1989, yields in these years may not be typical, but performance with below-average rain in 1990 helps give confidence that even with a sequence of low-rainfall years, the water harvesting systems which have been evolved would allow reasonable amounts of sorghum to be grown. There will be seasons, however, when rains are so slight that there is no significant runoff, and then no crop will be possible.

Apart from this, the main practical problem limiting crop production is pest control. Gardeners have been experiencing problems with beetles and the stem-borer *Chilo patellus*. Traditional methods of control were used, such as removing sorghum stover from the field, burning stalks, smoking and spreading ashes. There has also been a problem of sorghum plants growing well but failing to form heads, which may be due to rain coming at the critical head-stage in plant development, or else to the shoot fly *Atherigona*. On these matters, expert advice is being provided and efforts are directed to developing non-chemical treatment.

DIFFUSION AND EXPANSION

At the time of hand-over to self-management, the project was centred on three locations. Since then, discussions with groups along the lake shore have encouraged staff to suggest a fourth committee at Nachukwi, and by 1990, as already noted, this committee had improved some six gardens. This is, in a sense, a satellite committee to Loarengak. Satellite committees may also develop in the other two areas, perhaps to work with nomadic groups and open up a new dimension in local development. Rather refreshingly, however, for the first time, it will be people from the area who make the decision. New committees will stretch staff resources, but it is hoped that gaps will be filled by trained *ere* or *adakar* leaders.

To the west of Lokitaung Division, some quite separate work at Lopurr near Kakuma was initiated as a result of government concern that little progress was being made in rainwater harvesting in other parts of Turkana. The District Water Harvesting Officer (himself a Turkana) asked that we help establish a programme in the Kakuma area, where the Ngilukomong section have traditionally planted along the Tarac (Tarach) River. This fitted neatly with the Management Board's perception that more thought ought to be given to spreading water harvesting concepts to other Turkana.

By August 1989, a baseline survey had been completed for Kakuma, paid for by Oxfam (Fry, 1989). Much of the survey work was carried out by Peter Lomukat, who formerly worked with Oxfam's restocking project. In the early part of the year, Peter spent some time in the Loarengak and Lokitaung areas with a view to applying experience of this area in Kakuma Division. The report he wrote after visiting the lake-shore gardens (*shambas*) is of interest in showing awareness of some problems not mentioned so far, and staff at Lokitaung were clearly challenged by his comments. The visit was made after the unusually early rains in February 1989, when some bunds had been broken, but also, some early crops had been established. When the Loarengak extensionist Eyenae Ekaruan demonstrated the use of a line level, Peter reported as follows:

The line level, he said, needs about four people to operate: two holding the sticks, while a surveyor checks the bubble in the middle and the other person carries pegs or markers. The surveyor suspends the level from the centre of the string, which should be marked with a knot. He then instructs one person to move his stick up or down the slope till the bubble is levelled. A peg is placed to mark the point and the first operator puts a stick against the new peg ... and the process is repeated ... We did a little practice in pegging and measurement in Nakitokoonon in a *shamba* that was half-way complete, and Eyenae promised to show us how to do it in a new *shamba* ... His work as a designer was the most interesting to know about ...

The *shambas* in Loarengak will benefit if another rain is received soon, because right now some crops have begun reaching full height in those *shambas* that received enough water and for those that never had enough water are beginning to be slightly yellowish ...

Most *shambas* are only half-way cultivated. Maybe the owners lacked people to help them finish. Broken bunds are also one of the reasons that delayed some families. Otherwise all the *shambas* are doing well.

During its first year, the Kakuma project evolved a more broadly-based programme than the Lokitaung and Kachoda valley area had begun with, including the marketing of skins and the supply of some basic veterinary medicines. A committee was established and leaders were selected, and the project registered as a self-help group. By September 1990, there were thirty-four members, comprising fourteen men and twenty women, and the membership fee was KSh5/-.

Water harvesting work started from a local traditional practice of planting grass strips along the contours (and hence across the flow line of surface runoff). These strips were regarded partly as boundaries between gardens, but were seen to contribute to better water-spreading (Martin, 1990). Rainfall is higher at Kakuma than at Lokitaung, and it should be anticipated that optimum techniques will prove to be different. Grass strips may well provide sufficient runoff control without need for heavy bund construction. Either way, the Kakuma work has an advantage in starting from local people deciding what they want to do, rather than from a food-for-work programme directed from outside.

CEREAL STORES AND SKINS MARKETING

By the beginning of 1989, the project in the Loarengak, Kachoda and Kaalin locations, formerly known as the Turkana Water Harvesting Demonstration Project, was renamed the Lokitaung Pastoral Development Project. The new name was chosen because the project had developed broader objectives, with one aspect of its work now conceived in terms of 'food security' in the widest sense. This was seen to cover three areas of work: firstly, rainwater harvesting for improved cereal production; secondly, community stores, which were then newly opened; and lastly, livestock production and marketing, a possibility still to be developed, but one for which the Kakuma project may provide some stimulus.

The idea of developing food stores arose after local people had attended training sessions and workshops outside the area. Time spent in Samburu was particularly fruitful and staff saw at first hand local stores and skin marketing organized by groups of restocked herders. The idea appealed to project staff and was discussed in detail with *ere* representatives.

Although 1988 was to prove a year of above average rainfall, the preceding dry season had been long and hard, and there were large numbers of deaths in goat and sheepherds. As the numbers of pastoralists selling skins from these animals increased, traders forced skin prices down from KSh25/– to KSh20/–, then to half this, and finally down to KSh5 to 6/– per skin. Falling prices meant that some people had difficulty in buying food, and this prompted the idea that the lessons learned in Samburu should be immediately applied. As staff at Kaalin were the keenest, the idea was first tested there. Staff were given a grant of KSh3000/– for intervention buying of skins at KSh20/– per piece. Within a week the money was used, but fortunately, a local trader offered to buy the skins for KSh23/50 each, and so the fund was re-established.

Since then, staff and pastoralists in other locations have started similar schemes and prices were increased throughout the region. It needs to be said, however, that intervention buying would have been impossible without the approval of the Department of Hides and Skins at Lodwar. The officer in charge was particularly helpful in providing a letter authorizing the Kaalin committee to purchase skins. Due to proposed changes in legislation, permits were not then being granted, and the project could have been denied support (Cullis, 1988).

In sharp contrast to the rainwater harvesting work, this initiative required only start-up funds. Staff training was limited to simple book-keeping. The success of skin buying encouraged a more serious examination of other initiatives which could complement the rainwater harvesting work. The committees registered with the Ministry of Culture and Social Services as self-help groups so that they could gain official clearance for group activities and trading. In order to register, each committee had to elect a chairperson, treasurer and secretary. Both staff and non-staff were elected to these posts, which was a step forward in encouraging wider participation in management of the project. More recently, people have gained greater confidence, and more positions have been filled by non-staff members. These include the chairman and chairlady of each local committee.

Once the stores were properly registered as being run by self-help groups, it was possible to enrol group members on a formal basis. After some discussion, the committees decided to set a small annual membership subscription. By mid-1990, there was a total of 250 members at the three locations. The completed stores quickly proved to be of great value for the safe keeping of tools and seeds as well as for the skins and foodstuffs being traded. In the first year, some gardeners also deposited part of their sorghum harvest in the stores as previously mentioned, and in the second year, such was the demand that the Loarengak store was quickly filled to capacity with several tonnes of grain. Extra space is needed on a temporary basis to cater for this large post-harvest demand and suggestions have been made about construction of traditional buildings with additional waterproofing (Martin, 1990).

Where food sales are concerned, demand has tended to exceed supply, and during 1990, the Management Board had to release extra funds as a loan for purchase and transport of food stocks from Lodwar. All three stores were making profits and expected to have no difficulty in repaying the loan. The prices at which food is sold have tended to vary between locations, with the Loarengak committee having to meet higher transport costs as well as wishing to build up their funds. Prices are subject to Management Board approval, though in the longer-term, stores are expected to be self-sufficient in supporting their staff and paying for purchases.

The main items sold from the shops are maize, maize flour, sugar and tobacco, with sugar sales increasing particularly quickly (Martin, 1990). Although the sale of tobacco and sugar reflects local wishes, it is worth noting that these items do not contribute to food security, except by contributing to the overall viability of the stores. Increasing sugar consumption can sometimes be associated with a worsening of malnutrition in a community, because it provides only 'empty calories' – that is, nothing of nutritional value apart from energy. One of the paradoxes of 'development' in many parts of the world is that while traditional diets are usually well-balanced nutritionally – and this is very likely to be true in Turkana – the arrival of imported foods is often accompanied by a drop in nutritional standards and a need for health education and nutrition programmes.

Be that as it may, the progress of the stores in Lokitaung Division is considered to be contributing to food security by making maize available at reasonable prices and reducing exploitation by traders, also by providing secure storage for harvested crops, seeds, and agricultural equipment. The trade in skins links closely with the food security objectives in giving good prices for skins and hides, thereby giving pastoralists the means to buy food when they need to. Both the skins trade and the stores are of benefit to the wider pastoral community since the stores buy and sell to anyone in the area (not just members), although food may be reserved for members in times of shortage. If staff need to be convinced of the value of the work, they have only to listen to the very appreciative comments of people buying food and seed (Martin, 1990).

CONCLUSION:
DIALOGUE-BASED DEVELOPMENT

SAHEL-ZONE PROBLEMS

The African Sahel, with its low agro-ecological potential and periodic famines, its civil wars and inter-ethnic raiding, has the poorest regional economy in the world. Recently it has fallen still further behind other regions as economic growth rates have stagnated. Forecasters suggest that this trend is unlikely to change in the near future.

Within the Sahel, pastoralists and agro-pastoralists are particularly vulnerable because development policy, often supported by international funding, has failed to address local concerns. Development programmes have been fraught with difficulty for other reasons as well, and not only government schemes. A review of Oxfam projects in the region has noted that many are 'administratively demanding' and need to be backed by better research (Muir, 1987). By contrast, a small number of projects stand out as exceptionally positive. One of the best-known and most widely written about is concerned with the control of runoff for improved crop production in the Yatenga region of Burkina Faso (Harrison, 1987; Pacey and Cullis, 1986). More recently, it has been possible to see that the Lokitaung Pastoral Development Project in Turkana is in some ways comparable (Muir, 1987), not only in its use of a basic water-conservation technology, but also in being based on priorities identified by local people, technology adapted by them, decisions made by them, and an unhurried, non-authoritarian approach which ensures that pastoralists and gardeners are able to take full responsibility for the work.

One key point about the Turkana project is that local people have always had a considerable degree of control over what is done, with the success of self-management since 1988 as the logical outcome. Another crucial factor has been the focus on techniques and methods of an 'appropriate' type and scale, capable of being operated and maintained with local resources. For example, implements are made by local black-smiths' methods, not in a western-style workshop with imported equipment. Water-harvesting techniques have been on a scale commensurate with the size of traditional sorghum gardens, not the larger scale possible with spate irrigation. The result has been an expansion in the area of land that is cropped, and more reliable and sometimes bigger grain yields.

The role of food-for-work in this needs to be noted, however, since the earth-moving work involved in improving gardens requires a massive labour input, to the extent that it is found to be very demanding by most families – particularly the poorer pastoralists' families, and women-headed households. For such people, some kind of food subsidy or payment for work has been very important. The rate of garden improvement slowed when TRP failed to deliver most of the maize owed as food-for-work payment during 1989, and the question which has to be tackled now is how low-level support can continue to be provided for garden improvement. It may be that a 'traditional loan fund' of credits to be cashed for grain at the stores and repaid from subsequent harvests can be devised. In addition, the increased use of animal draught is helping the situation by reducing labour requirements. Other options include moving slowly away from bund construction to deal with other constraints on production, such as seed supply, improved hand tools, and better pest control.

Exploring such options through discussion within local pastoral institutions means that in ten years' time the work may be entirely different from what we see today, just as today's project looks completely different from what was originally proposed by Oxfam's consultants. One way of characterizing the change which occurred is to say that while the consultants envisaged a 'demonstration project', there was an early move away from this concept towards a more 'participatory' approach for developing the water-harvesting systems. This process was not a simple one of discovery followed by action, but rather one of planning, action, reflection, and further planning, constantly repeated throughout the life of the project. Learning from mistakes (such as building bunds of unstable clay soil), and learning from people's reactions (e.g. to draught animals, or the availability of grain stores), became an important part of the process. After some initially encouraging results, it became necessary to consider how to institutionalize this process. In conjunction with the advisory committee of elders, it was agreed to adopt the *ere* or home area (to which families return in the wet season) as the base from which to build. Staff members, who were themselves ordinary ex-herders from the area, identified *ere* where work could be started, and held meetings with the *ere* elders both to learn from reactions to the techniques proposed, and to encourage them to use food-for-work to improve their gardens (or develop gardens in places where none existed).

THE PROCESS OF DIALOGUE

The difference between development based on demonstration and development involving participation merits further comment because there are implications both for understanding the necessary social process and for understanding how technology evolves.

The idea of a 'demonstration project' such as the one inaugurated in Turkana during January 1985 (Chapter 2) tends to imply that people with expertise (usually from outside the area) are showing something to other people who lack expertise, and who are expected simply to copy what they learn. Many projects which attempt to 'transfer' alien technology to 'developing countries' are based on such assumptions. Most go badly wrong because the assumptions are mistaken. Local people *do* have expertise and the expert or consultant ignores this at his peril. No matter how sophisticated the alien technology being introduced, local people will recognize social and environmental implications of its use which outside experts tend to miss. Where rainwater harvesting is concerned, there is the additional fact that the Turkana people have long experience of related techniques, and therefore have 'technical' knowledge of their own.

Being aware of at least some of this, Adrian organized discussion groups at an early stage in the 'demonstration project', and in so doing made it more 'participatory' and less of a demonstration. The discussions were intended to enable him to understand the pastoral way of life, to appreciate the role of local institutions, and to learn how traditional decision-making processes worked. They revealed that many elders felt frustrated and alienated from 'development' as they had so far experienced it. The discussions also enabled him to learn about the local economy, lifestyle and environment, and were a constant source of interest and stimulus.

To understand this process better, however, a stronger word is necessary to describe it – 'participation' sounds too much like a concession made by powerful outsiders, rather than an essential process involving real exchanges, commitments to other people, practical experiments and mutual learning. One view of how people come to accept new technology, then develop and innovate it, is that there is always a dialectical process or *dialogue* (Pacey, 1990). But 'dialogue' in this sense is not just a matter of talking. It also involves hands-on interaction with the techniques concerned, including trial-and-error experiment, responses to environmental impacts, and reactions to organizational and management problems. Processes of this sort may often be implied when 'participation' is referred to, but the word is often also used in a much weaker sense, as if participation were an optional extra.

To appreciate how a dialogue is possible between people with a background of western science and technology, and agro-pastoral people in the Sahel, one ought to be aware of the different kinds of knowledge characteristic of the two cultures, and the strengths and weaknesses of each. The emphasis of western science is on the discovery of basic principles which are universally valid. Thus when Israeli researchers elucidated the ancient runoff farming techniques used in the Negev Desert, there was a temptation to look for the general principles involved, and assume that they were applicable in all the drier areas of the world –

including many areas in Africa (Oxfam, 1974, section 12). Thus there was a failure to anticipate how sensitive the Israeli techniques would prove in relation to the 'ecological particulars' of different regions and environments. In the Sahel, attempts were made to copy the Negev Desert methods in both the Yatenga project mentioned previously and in Turkana, but with only limited success. Effective water harvesting only became possible in these areas when more attention was paid to local knowledge of agriculture and water conservation techniques and also to local institutions governing the use of land and the organization of work. It was the interaction of this local information with new ideas at every level, from informal discussion to practical experiment, which constituted the dialogue from which new approaches emerged.

The strength of local knowledge of the environment is nearly always related to keen observation and experienced discrimination. One example in Turkana is discrimination between different sorghum varieties and selection of the remarkable quick-growing 62-day variety. Another example is the distinction Turkana gardeners draw between 'red water' and 'brown water' in discussing runoff flows onto their land (Chapter 5). They implicitly recognize that the silt and organic-matter load (and also the flow volume) of brown water is better for their gardens.

But just as western science may have weaknesses resulting from over-confidence in its assumptions about universal application, so local knowledge has limitations also, the most obvious of which is the lack of a framework of quantification and abstract principle. It is significant that in both the Turkana and Yatenga projects, the one facet of western expertise which made the biggest difference to people's ability to control new technology was a skill usually dependent on mathematical reasoning – the technique for surveying levels. In both projects, the introduction of a levelling method which could be used without numeracy skills had a markedly liberating effect, opening the door to a confidence that all other aspects of the technology could be mastered (Chapter 4). In Yatenga, a water tube level was used, in contrast to the line level in Turkana, but the psycho-social impact was the same in both cases. Overcoming the obstacle which lack of numeracy might otherwise have entailed gave a boost to people's confidence in using all aspects of the improved technology which had become available.

DIMENSIONS OF DIALOGUE

One aspect of dialogue, then, is a sharing of knowledge. But the dialogue which took place in Turkana was also a matter of institutions, relationships and individual awareness. There were personal factors involved also, which have been expressed to a small degree by the way this book is written, with separate paragraphs set apart from the main body of the text where different people describe their own experience.

Some important points about how personal factors may influence a 'project dialogue' can best be made if I, Arnold Pacey, step outside my role as editor for a moment to say something independently of Adrian Cullis, the author. My experience of editing the writings of development workers over many years (Oxfam, 1974; Pacey and Payne, 1983) has made me aware that the commitment of individuals can add an extra dimension to the kind of dialogue which is possible. I have noticed three particular aspects of this:

Firstly, the most effective projects are often those which allow the work and its goals to evolve over several years, resisting demands that tangible results be achieved by fixed target dates. In the Turkana work, an occasion when pressure for results threatened the process of dialogue – and threatened Adrian's position also – was described in Chapter 4.

Secondly, many instances of this sort of unhurried dialogue have been made possible because an individual development worker has had a long-standing connection with a particular community, giving time for personal relationships to grow and learning experience to be digested.

In Turkana, Adrian was associated with water harvesting work from late 1979, five years before the main project described here began. On returning to the area in late 1984, he was able to pick up the threads of old friendships. The development worker's willingness to learn is also important, and it is striking that Adrian deliberately involved himself in this, not only by organizing discussions, but also by acquiring his own flock of goats (Chapter 4), by working in gardens to see exactly how seeds were planted, and by undergoing leadership training along with the project staff (Chapter 6).

Thirdly, I have noted that in the histories of 'successful' projects, there are nearly always false starts or episodes of apparently total failure, but these function as means of learning. In Turkana, mistakes made in the Salvation Army's water harvesting project and by TRP were part of this learning process. Mistakes are a *necessary* part of dialogue-based development, and in this context, the Salvation Army work, TRP and the Oxfam-backed project should be looked at together as parts of a single process.

The implications of this may seem unrealistic. Not every development worker can spend ten years with a project. But one can perhaps gain in understanding by trying to appreciate why a ten-year connection has been important in this instance. The main lessons are for the development agencies to learn: they can make longer-term commitments; they can be less pressing with demands for quick results; and they can do more to achieve continuity even if no individual stays with a project as long as indicated here.

Stating some of these points in terms of personal experience is also
important in that there is always a degree of tension between personal,
informal approaches and the need for formal management and systematic
research. In this work, such tension was experienced in collecting information about Turkana's life and the local economy. At first, this was done in
an informal, unstructured way. Later, as described in Chapter 5, a more
organized data collecting or monitoring system was set up, with the help of
a social scientist, Adrienne Martin. A research project focusing specifically
on the role of Turkana women was also established (Watson, 1986; 1988).
However, the informal approach was maintained alongside quantitative
data collection as it provided many extra insights (Cullis and Watson,
1989). Parallel use of informal and 'scientific' approaches may always be
important for dialogue-based development, the informal part of the work
being vital for ensuring that what takes place is really an exchange of ideas,
and that local people have an opportunity to learn about the development
workers as well as vice versa.

One general implication of all this is the importance of thinking in terms
of *processes* rather than *projects*. The distinction is that projects tend to be
limited in time, and to have defined goals. They are then judged in terms of
success or failure according to whether they reach those goals. By contrast,
a process goes on indefinitely into the future. It may have encouraging
phases as well as less promising interludes, but never reaches a final success
– or failure. Moreover, a process can spread beyond its initial 'project area'
to involve people in other places.

For example, the Lokitaung Pastoral Development Project (LPDP) has
had three years of good harvests (1988, 89 and 90) during which stores have
been established and the management structure has worked well, as
described in Chapter 6. This has been very encouraging, but it is only a
stage in a longer-term process, not a final success. Certainly, it has been
sufficiently encouraging to stimulate work in other places, most notably in
the sister project at Kakuma. Influence has also spread through the wider
network of Oxfam-supported pastoral projects in the region, and even into
Kenya government planning documents (ASAL, 1990).

However, it would be a mistake to isolate LPDP achievements in 1988–
90 as 'success' and dismiss everything else in the area as 'failure'. The
Salvation Army and TRP made mistakes which had to be made if water
harvesting was ever to develop. LPDP was a major beneficiary of what was
learned this way. Certainly, TRP made its mistakes on an unusually grand
scale, but this was the result of pressure to provide food-for-work for
thousands of temporarily destitute people. Thus TRP should be given
credit for fulfilling its primary famine relief role and saving many lives. At
the same time, TRP experience of wrongly-sized catchments, spillway
problems and breached bunds contributed to the design philosophy with
which LPDP now operates (Chapter 5).

The process which has been under way within and around LPDP will, of course, continue into the future, and we must expect changes in direction, especially if environmental or economic conditions alter greatly. The most important thing is that a group of people in Turkana have gone through a learning process whose impact has clearly been positive. Whether bunds continue to be built is less important than the fact that this process has begun and promises to empower people to cope better with economic pressures and their harsh environment.

REFERENCES

Adams, M. (1986), 'Merging relief and development: the case of Turkana', *Development Policy Review*, Vol. 4, pp.314–24.

ASAL, (1990), Republic of Kenya ASAL Policy Development 27 September, Ministry of Reclamation and Development of Arid, Semi-Arid Areas and Wastelands.

Barasa, S. (1986), Notes on implements for use with draught animals. Turkana Water Harvesting and Draught Animal Demonstration Project.

Barrow, E. G. C. (1987), 'Extension and learning; examples from the Pokot and Turkana'. IDS Workshop on Farmers and Agricultural Research, University of Sussex.

Barrow, E. G. C. (1988), *Trees and pastoralists: the case of the Pokot and Turkana*, ODI Social Forestry Network Paper 6b.

Boogaard, R. van der, (1986), *Nutritional status of under-5 children in Turkana District*. TRP/TDSU Ministry of Energy and Regional Development.

Boserup, E. (1981), *Population and Technology*. Blackwell, Oxford.

Brainard, J. M. (1981), *Herders to Farmers: the effect of settlement on the demography of the Turkana population of Kenya*. Ph.D. dissertation, State University of New York, Binghampton.

Broch-due, V. (1983), 'Women at the backstage of development: the negative impact on the project realization by neglecting the crucial roles of Turkana women from Katilu Irrigation Scheme'. Rome: FAO consultant's report, GCP/KEN/048/NOD.

Broch-due, V., and Storas, F. (1980), *Pastoral women in the process of social change*. NORAD, Lodwar.

Critchley, W. and Siegart, K. (1991) *Water Harvesting: A manual for the design and construction of water harvesting schemes for plant production*, FAO, Rome, AGL/MISC/17/91

Cullis, A. (1981), 'A study of runoff farming in Turkana District of northwest Kenya'. Report on a field project for Salford University.

Cullis, A. (1988), 'Turkana Water Harvesting Project Handing-over Report', December 1988. ITDG, Rugby.

Cullis, A. (1989), 'Dryland Food Security Programme visit to Kenya, 5 June–2 July'. ITDG, Rugby.

Cullis, A., and Watson, C. (1989), 'Working with pastoralists: a case study from Turkana'. In proceedings of the Agriculture and Fisheries Sector Forum on Socially Appropriate Technologies, Coventry, ITDG, Rugby.

EEC, (1980), Recurring famine in Turkana: proposed strategy for alleviation. EEC (unpublished report), Brussels.

Evenari, M., Shanan, L., and Tadmor, N. (1982), *The Negev: the Challenge of a Desert.* Cambridge (Mass.), Harvard University Press, 2nd edition (1st ed 1971).

Fallon, L. F. (1963), Water spreading in Turkana. USAID Mission to Kenya, Nairobi.

Finkel, M. (1985), *Draft Turkana Water Harvesting Manual,* based on a water harvesting course carried out for the Ministry of Agriculture, funded by NORAD, ed. C. Erukudi and E. Barrow (revised ed. 1987).

Fry, P. (1989), 'Needs assessment survey, Kakuma Divison'. Report for Oxfam.

Grandin, B. E. (1988), *Wealth ranking in smallholder communities: a field manual,* IT Publications, London.

Gulliver, P. H. (1951), *A preliminary survey of the Turkana: a report compiled for the Government of Kenya,* School of African Studies, University of Cape Town. (1963 edition also referred to).

Gulliver, P. H. (1955), *The Family Herds: a study of two pastoral tribes in East Africa.* Routledge and Kegan Paul, London.

Gulliver, P. H. (undated) 'Nomadism among the pastoral Turkana: its natural and social environment'. *Nkanka,* 4.

Harrison, P. (1987), *The Greening of Africa,* Paladin Grafton Books, London.

Hartley, B. (1984), 'Demonstration of water management systems in Turkana': Oxfam project proposal, also report for Government of Kenya, TRP, and World Food Programme.

Henrikson, G. (1974), 'Problems of development in Turkana: economic growth and ecological balance'. Institute of Social Anthropology, University of Bergen, Occasional Paper No. 11.

Hillman, F. (1980), 'Water harvesting in Turkana District, Kenya'. ODI Pastoral Network Paper/Social Forestry Network Paper 10d.

Hogg, R. S. (1982), 'Destitution and development: the Turkana of north-west Kenya'. *Disasters.* 6 (3), p. 164–8.

Hogg, R. S. (1982), 'Destitution and development: a strategy for Turkana'. Department of Anthropology, University of Manchester.

Hogg, R. S. (1986), 'Water harvesting in semi-arid Kenya: opportunities and constraints'. Report for Oxfam.

Hogg, R. S. (1987), 'Building pastoral institutions: a strategy for Turkana District'. Report for Oxfam.

Hogg, R. S. (1988), 'Water harvesting and agricultural production in semi-arid Kenya'. *Development and Change,* Vol. 19, pp.69–87.

Hope, A. and Timmel, S. (1984), *Training for Transformation*, Mamo Press, Zimbabwe.Lamphear, J. (1976), Aspects of Turkana leadership during the era of primary resistance, *Journal of African History*, Vol. 17. 2, pp. 225–43.

McCabe, J. T. (1985), 'Livestock management amongst the Turkana: a social and ecological analysis of herding in an East African pastoral population'. Ph.D. thesis, State University of New York, Binghampton.

McCabe, J. T., and Ellis, J. (1987), 'Beating the odds in arid Africa'. *Natural History*, 1/87.

McCabe, J. T., and Fry, P. (1986), 'A Comparison of Two Survey Methods on Pastoral Turkana Migration Patterns and the Implications for Development Planning', Paper 22b, Pastoral Development Network, ODI.London.

Martin, A. M. (1986), 'Monitoring the Turkana water harvesting project'. Report for Oxfam and ITDG.

Martin, A. M. (1990), 'Lokitaung Pastoral Development Project: a review'. Report for Oxfam and ITDG.

Martin, A. M., and Gibbon, D. (1987), 'Turkana Water Harvesting Project: a review for Oxfam and ITDG'.

Ministry of Energy and Regional Development, 1985. 'Turkana District resources survey 1982–4'. Report by Ecosystems Limited.

Ministry of Finance and Planning, (1983), Turkana District Development Plan, 1984–8.

Morgan, W. T. W. (1974), 'The South Turkana Expedition: sorghum gardens in south Turkana', *Geographical Journal, 140* (1974), pp.80–98.

Morgan, W. T. W. (1980), 'Vernacular plant names and utilization of plant species in Turkana'. Department of Geography, Durham University.

Moris, J. (1987), 'Irrigation as a privileged solution in African development'. *Development Policy Review*, Vol. 5, pp.99–123.

Moris, J. (1988), 'Failing to cope with drought: the plight of Africa's ex-pastoralists'. *Development Policy Review*, Vol. 6, pp.269–94.

Muir, A. (1987), Oxfam Discussion Paper.

NORAD, (1979), Report on the development of the Turkana District, Kenya. Norwegian Agency for International Development, Oslo.

NORAD, (1990), Turkana Rural Development Programme: plan of operations. Norwegian Agency for International Development, Oslo.

Norconsult, A. S. (1978), Report of a fact-finding mission to Kalakol, Turkana District: environmental problems associated with economic development in a remote area of Kenya. Report for NORAD, Oslo.

ODI, (1985), Turkana District development strategy and programme, 1985/86–1987/88. Overseas Development Institute, London.

Oxfam, 1963 to date. Project files, KEN 84, KEN 109, KEN 197, etc.

Oxfam, (1974), *Field Directors' Handbook*, 1st edition, compiled by A. Pacey; (2nd ed, A. Pacey, 1976; revised ed, J. Alderson, 1980; Oxford University Press edition, B. Pratt and J. Boyden, 1985).

Oxfam ILED/RITA/ALIN, *Looking After Our Land: Soil and Water Conservation in Dryland Africa* video, Oxfam Publications, Oxford.

Pacey, A. (1990), *Technology in World Civilization*, Blackwell, Oxford.

Pacey, A., and Cullis, A. (1986), *Rainwater Harvesting*. IT Publications, London.

Pacey, A., and Payne, P., (eds), (1983), *Agricultural Development and Nutrition*, Hutchinson, London.

Powell, W. I. (1982), 'Irrigation in arid regions: Kenya. Water harvesting for crops'. Consultant's report, NORAD, and FAO document GCP/KEN/048/NOD.

Pratt, D., and Gwynne, M. (eds), (1977), *Rangeland Management and Ecology in East Africa*. Hodder and Stoughton, London.

Reij, C. (1987), Soil and water conservation in Sub-Saharan Africa: the need for a bottom-up approach. Paper for Benin conference.

Richards, P. (1985), *Indigenous Agricultural Revolution: Ecology and Food Production in West Africa*. Hutchinson, London.

Salvation Army, 1978 to date. Project files.

Sandford, R. (1984), Turkana District Livestock Development Plan. Report for Oxfam.

Schlee, G. (1985), 'Inter-ethnic clan identities among Cushitic-speaking pastoralists'. *Africa*, Vol. 55 (1), pp.17–37.

Schwartz, S., Schwartz, H., and van Dongen, P. (1985), Turkana Rehabilitation Programme, project evaluation study, final report for the Government of Kenya.

Sobania, N. (1975), 'The historical tradition of the people of the eastern Lake Turkana basin, c.1840–1925'. Ph.D. thesis, School of Oriental and African Studies, University of London.

Sorbo, G., Skjonsberg, E., and Okumu, J., (1988), NORAD in Turkana: A review of the Turkana Rural Development Programme. Draft report, Bergen, Norway.

Stern, P. (1979), *Small-scale Irrigation*. IT Publications, London.

Storas, F. (in press), 'Intention or implication: the effects of Turkana social organization on ecological balances'. In P. Baxter (ed.), *Property, poverty and people: changing rights in property and problems of pastoral development*, Dept. of Social Anthropology and International Development Centre, University of Manchester, UK.

Swift, J. (1981), 'Labour and subsistence in a pastoral economy'. In R. Chambers, R. Longhurst and A. Pacey (eds), *Seasonal Dimensions to Rural Poverty*, Frances Pinter, London.

Swift, J. (1985), Memorandum to Oxfam.

Swift, J. (1988), Drought planning in Turkana. Report for Oxfam.

Swift, J., Cullis, A., and Watson, C. (1986), 'Turkana Water Harvesting and Draught Animal Demonstration Project: mid-term review', Oxfam and ITDG.

Thomas, R. G. (1982), (a) Back-to-office report to PDR Yemen; (b) 'Ancient spate irrigation in Wadi Beihan', Yemen. FAO, Rome.

TRP, 1982–3. Project files.

Turton, D. (1989). Personal Communication.

UNICEF, (1985), *Within Human Reach: a Future for Africa's Children*. United Nations Children's Fund, New York and Geneva.

Watson, C. (1986), 'The Women of Manalongoria'. Provisional report to Oxfam and ITDG.

Watson, C. (1988), 'The development needs of Turkana women'. Report for Oxfam and the Public Law Institute, Nairobi.

Watson, C., with Lobuin A. (1990), 'Turkana Water Harvesting Project: lake shore study'. ITDG Dryland Food Security Programme.